Applied Artificial Intelligence in Data Science, Cloud Computing and IoT Frameworks

(Volume 3)

AI and ML Solutions Driving Modern Farming and Urban Innovation

Edited by

Suneeta Satpathy
Center For Cybersecurity, SoA Deemed to be University, Bhubaneswar, Odisha, India

Sachi Nandan Mohanty
School of Computer Science & Engineering, VIT-AP University, Andhra Pradesh, India

&

Subhendu Kumar Pani
Computer Science & Engineering Krupajal Engineering College BPUT, Odisha, India

Applied Artificial Intelligence in Data Science, Cloud Computing and IoT Frameworks

(Volume 3)

AI and ML Solutions Driving Modern Farming and Urban Innovation

Editors: Suneeta Satpathy, Sachi Nandan Mohanty and Subhendu Kumar Pani

ISSN (Online): 3029-2247

ISSN (Print): 3029-2255

ISBN (Online): 979-8-89881-210-2

ISBN (Print): 979-8-89881-211-9

ISBN (Paperback): 979-8-89881-212-6

Published by Bentham Science Publishers Pte. Ltd. Singapore, in collaboration with Eureka Conferences, USA.

First published in 2025.

Bentham Science Publishers Pte. Ltd.
No. 9 Raffles Place
Office No. 26-01
Singapore 048619
Singapore
Email: subscriptions@benthamscience.net

CONTENTS

FOREWORD

In an era defined by technological convergence, Applied Artificial Intelligence in Data Science, Cloud Computing and IoT Frameworks: Volume 2 – AI and ML Solutions Driving Modern Farming and Urban Innovation emerges as a timely and visionary contribution to the global discourse on intelligent transformation. This volume arrives at a moment when Artificial Intelligence (AI) and Machine Learning (ML) have evolved beyond research laboratories to become foundational forces shaping industries, economies, and societies. Their integration with Cloud Computing and the Internet of Things (IoT) is driving unprecedented innovation, enabling sustainable solutions, and redefining the way humanity interacts with its environment.

The agricultural and urban sectors, in particular, lie at the heart of this transformation. Confronted by challenges such as food insecurity, climate change, urban congestion, and resource inefficiency, these domains demand solutions that blend intelligence with sustainability. AI-powered analytics, when coupled with the scalability of cloud infrastructures and the real-time responsiveness of IoT ecosystems, provide a powerful framework for addressing these issues. From precision agriculture that optimizes yield while conserving resources, to smart city systems that enhance mobility and reduce emissions, the innovations explored within this volume exemplify how data-driven intelligence can catalyze equitable and sustainable progress.

Distinguished by its synthesis of theoretical depth and practical insight, this book bridges the gap between conceptual understanding and real-world implementation. Each chapter presents rigorous analyses alongside applied frameworks, case studies, and experimental outcomes that illuminate the tangible impact of AI and ML technologies across rural and urban landscapes. By integrating cross-disciplinary perspectives, the volume offers researchers, practitioners, and policymakers a comprehensive guide to advancing smarter, greener, and more resilient ecosystems.

Beyond its technical merit, this volume stands as a reflection of collective human ingenuity directed toward the greater good. It envisions a future where innovation serves not only as a tool of progress but as a means of harmonizing technological advancement with environmental and societal well-being. The editors and contributors have assembled a body of work that captures both the promise and responsibility of intelligent systems in shaping the next era of sustainable development.

Shu Hu
School of Applied and Creative Computing
Purdue Machine Learning and Media Forensics (M2) Lab
Purdue University
West Lafayette, Indiana

PREFACE

The rapid evolution of Artificial Intelligence (AI) and Machine Learning (ML) is reshaping the way societies address some of their most pressing challenges. This volume, ***AI and ML Solutions Driving Modern Farming and Urban Innovation***, brings together a diverse collection of research contributions that highlight how intelligent technologies are transforming both agricultural practices and urban living.

Modern farming today demands precision, efficiency, and sustainability. AI and ML are enabling farmers to optimize crop yields, predict diseases, and manage resources with unprecedented accuracy. Similarly, cities around the world are turning to intelligent systems to tackle complex issues such as waste management, water quality monitoring, traffic congestion, and environmental sustainability. Together, these innovations demonstrate the potential of AI and ML to create smarter, healthier, and more resilient communities.

This volume serves as a platform for scholars, practitioners, and innovators to share advancements, methodologies, and applications that bridge the gap between research and real-world deployment. By showcasing interdisciplinary solutions, it emphasizes the societal, local, and national impacts of intelligent technologies in agriculture and smart city ecosystems.

We hope this collection inspires further exploration and collaboration toward building sustainable futures powered by AI and ML.

Suneeta Satpathy
Center For Cybersecurity, SoA Deemed to be University
Bhubaneswar, Odisha, India

Sachi Nandan Mohanty
School of Computer Science & Engineering
VIT-AP University, Andhra Pradesh, India

&

Subhendu Kumar Pani
Computer Science & Engineering, Krupajal Engineering College BPUT
Odisha, India

List of Contributors

Alladi Vennela	School of Computer Science and Engineering (SCOPE), VIT-AP University, Amaravati, Andhra Pradesh, 522237, India
A.L. Kalyani	Department of CSE, Shri Vishnu Engineering College for Women(A), Bhimavaram, Andhra Pradesh, 534202, India
A. Ezil Sam Leni	Centre of Excellence in Computer Vision, Department of Computer Science and Engineering, School of Advanced Computing, Alliance University, Bangalore, Karnataka, 562106, India
Aanchal Chaudhary	School of Computer Engineering, KIIT (Deemed to Be) University, Bhubaneswar, 751024, Odisha, India
Ayush Yadav	School of Computer Engineering, KIIT (Deemed to Be) University, Bhubaneswar, 751024, Odisha, India
Brehanu Borji	Department of Management, Hawasa University, Hawasa, Ethiopia
B.V. Chandrahaas	School of Computer Science & Engineering, VIT-AP University, Amaravati, Andhra Pradesh, India
Bhagavathi Santos Kumar	Department of Computer Science and Engineering, School of Advanced Computing, Alliance University, Bangalore, Karnataka, 562106, India
Chetan J. Shelke	Centre of Excellence in Computer Vision, Department of Computer Science and Engineering, School of Advanced Computing, Alliance University, Bangalore, Karnataka, 562106, India
Chalchisa Amentie	Department of Management, Ethiopian Civil Service University, Addis Ababa, Ethiopia
C. Sai Janakiram	School of Computer Science & Engineering, VIT-AP University, Amaravati, Andhra Pradesh, India
Durai Selvaraj	Department of CSE, Manipal Institute of Technology, Manipal Academy of Higher Education, Manipal, Karnataka, India
Ganesh Reddy Karri	SCOPE, VIT-AP University, Andhra Pradesh, India
Harsh Kumar Pandey	School of Computer Engineering, KIIT (Deemed to Be) University, Bhubaneswar, 751024, Odisha, India
Hitesh Mohapatra	School of Computer Engineering, KIIT (Deemed to Be) University, Bhubaneswar, 751024, Odisha, India
J.L.B. Krishna	School of Computer Science & Engineering, VIT-AP University, Amaravati, Andhra Pradesh, India
J. Sheela	School of Computer Science & Engineering, VIT-AP University, Amaravati, Andhra Pradesh, India
K. Varada Rajkumar	Department of CSE (AI&ML), MLR Institute of Technology, Telangana, India
Kiran Sree Pokkuluri	Department of CSE, Shri Vishnu Engineering College for Women, Bhimavaram, India
Lalitha	School of Advanced Sciences, VIT-AP University, Andhra Pradesh, India

Manoj B. Chandak	Department of Computer Science and Engineering, Ramdeo Baba University, Nagpur, Maharashtra, India
M. Vaishali	Department of Computer Science and Engineering, Maharishi Markandeshwar Deemed to be University, Ambala, India
M. Monika	Department of Information Technology, Dwarkadas J. Sanghvi College of Engineering, Navi Mumbai, India
Mershaye Birhane	Department of Management, Bule Hora University, Bule Hora, Ethiopia
Mohamed Iqbal M.	Department of Information Technology, Sri Sivasubramaniya Nadar College of Engineering, Chennai, 603110, India
M. Vaishali	Department of Computer Science and Engineering, Maharishi Markandeshwar Deemed to be University, Ambala, India
M. Monika	Department of Information Technology, Dwarkadas J Sanghvi College of Engineering, Navi Mumbai, India
M. Atish	Department of Computer Science and Engineering, BMS Institute of Technology and Management, Banglore, India
M. Rushane	School of Computer Science & Engineering, VIT-AP University, Amaravati, Andhra Pradesh, India
Nishant Singh	School of Computer Engineering, KIIT (Deemed to Be) University, Bhubaneswar, 751024, Odisha, India
Prakruthi Ganiga	Centre of Excellence in Computer Vision, Department of Computer Science and Engineering, School of Advanced Computing, Alliance University, Bangalore, Karnataka, 562106, India
Pelluru Pavan Kumar Reddy	SCOPE, VIT-AP University, Andhra Pradesh, India
Rathnakar Achary	Centre of Excellence in Computer Vision, Department of Computer Science and Engineering, School of Advanced Computing, Alliance University, Bangalore, Karnataka, 562106, India
Ramalakshmi K.	Centre of Excellence in Computer Vision, Department of Computer Science and Engineering, School of Advanced Computing, Alliance University, Bangalore, Karnataka, 562106, India
Ramith Yadav S.	SCOPE, VIT-AP University, Andhra Pradesh, India
Rushi Patel	Department of Electronic and Telecommunication Engineering, Indian Institute of Technology - IIT, Kharagpur, 721302, West Bengal, India
Radha Mohan Pattanayak	School of Computer Science and Engineering (SCOPE), VIT-AP University, Amaravati, Andhra Pradesh, 522237, India
Shweta Bhelonde	Department of Computer Science and Engineering, Ramdeo Baba University, Nagpur, Maharashtra, India
Sudheer Mangalampalli	Department of CSE, Manipal Institute of Technology, Manipal Academy of Higher Education, Manipal, India
Safiya Begam	B. S. Abdur Rahman Crescent Institute of Science and Technology, Chennai, Tamil Nadu, India

Sharon Priya	B. S. Abdur Rahman Crescent Institute of Science and Technology, Chennai, Tamil Nadu, India
S. Rajarshi	School of Advanced Sciences, VIT-AP University, Andhra Pradesh, India
Shashi Kant	Department of Management, Bule Hora University, Bule Hora, Ethiopia
S. Divya Meena	School of Computer Science & Engineering, VIT-AP University, Amaravati, Andhra Pradesh, India
Sujithra Thandapani	Department of CSE, Manipal Institute of Technology, Manipal Academy of Higher Education, Manipal, Karnataka, India
Swati Shukla	School of Electronics Engineering (SENSE), VIT-AP University, Amaravati, 522237, Andhra Pradesh, India
Smital Anandrao Patil	School of Computer Science, KIT's College of Engineering, Kolhapur, 416234, Maharashtra, India
Vaibhav Kumar Singh	School of Computer Engineering, KIIT (Deemed to Be) University, Bhubaneswar, 751024, Odisha, India
V.C. Bharathi	School of Computer Science & Engineering, VIT-AP University, Amaravati, Andhra Pradesh, India
Virendra Kumar Shrivastava	Department of Computer Science and Engineering, School of Advanced Computing, Alliance University, Bangalore, Karnataka, 562106, India
W. Aisha Banu	B. S. Abdur Rahman Crescent Institute of Science and Technology, Chennai, Tamil Nadu, India
Yeswanth Satya Prasad	School of Computer Science and Engineering (SCOPE), VIT University, Vellore, 632014, Tamil Nadu, India
Yagnesh Challagundla	School of Computer Science and Engineering (SCOPE), VIT-AP University, Amaravati, 522237, Andhra Pradesh, India

Section 1
Modern Farming with Advanced Intelligence

CHAPTER 1

A Review on Paddy Plant Disease Detection Using Various Machine Learning and Deep Learning Algorithms

Pokkuluri Kiran Sree[1,*] and **A.L. Kalyani**[1]

[1] *Department of CSE, Shri Vishnu Engineering College for Women(A), Bhimavaram-534202, Andhra Pradesh, India*

Abstract: Early detection of plant diseases is critical in the agricultural sector to reduce the loss caused by the diseases. Paddy crop diseases impact yields in various ways, given that the crop is cultivated globally. The specific disease manifestations and severity, however, often vary with climatic and geographical conditions. The method used to treat the disease in one area may not be suitable in other areas. The manual detection of diseases by farmers is a very costly, time-consuming, and tedious process. To avoid these issues, researchers have developed various automatic methods. Most models utilize image processing and machine learning, or deep learning algorithms, to achieve the designated output. This study analyzed various algorithms used to automate the plant disease detection process, examining their advantages and disadvantages.

Keywords: Paddy crop diseases, Arrange alphabetically, Machine learning, Deep learning, Image processing.

INTRODUCTION

Agriculture is the foundational pillar on which many human societies have lived for thousands of years. Agriculture is the primary source of nourishment for billions of people worldwide by providing essential foods. Agriculture impacts various aspects of human life, including the economy, employment, and environment, particularly in rural areas. The agricultural sector also includes agribusiness, food processing, distribution, and farming. Out of all the crops growing worldwide, more than 50% of the world's population depends on paddy as a primary food crop. Paddy stands at the forefront of all crops, ensuring food security and providing nutritional food rich in calories, essential for the global population. Paddy crops are the primary income source for most farmers in the countryside, and they also support various industries along the agricultural supply

* **Corresponding author Pokkuluri Kiran Sree:** Department of CSE, Shri Vishnu Engineering College for Women(A), Bhimavaram-534202, Andhra Pradesh, India; E-mail: drkiransree@gmail.com

Suneeta Satpathy, Sachi Nandan Mohanty & Subhendu Kumar Pani (Eds.)

chain. Paddy also holds cultural and traditional significance for worldwide communities. Paddy is a source of Carbohydrates and nutrients are the basic food for most of the global population. Paddy plants can grow in various environments, including lowland flooded areas and highland regions. Additionally, it serves as a good food source for diverse geographical and climatic conditions.

Despite the importance of paddy crops, paddy diseases have a significant impact on production and farmers' lives. It is essential to understand the disease's causes, diagnose its symptoms, and provide better treatment methods in the early stages of the crop. Manual diagnosis of the disease by the farmer's naked eye may not be sufficient, as the farmer must inspect every plant individually, which can be time-consuming and require a significant workforce. This may even lead to the loss of an entire crop if not properly taken care of by the farmer. All these manual processes are being automated using machine learning algorithms, deep learning algorithms, and other advanced technologies. Almost all of the proposed models failed to detect the diseases in the main field. All the models were trained and tested using the controlled images with transparent backgrounds and single leaves. These models should be trained on field images, enabling farmers to utilize them directly to save time and effort and achieve more accurate disease detection results.

Paddy Crop Diseases: Over 50% of the world's population gets food from paddy crops. Paddy crops will be affected by various diseases due to climatic conditions. The diseases vary depending on the climate and geographical conditions. The major diseases that affect the paddy crop are discussed below. Such diseases are mainly classified into:

Fungal Diseases

Blast:

Also called Pyricularia oryzae. The fungus will attack the crop at all growth stages in this disease. The disease symptoms will appear on leaves, nodes, and glumes. Blast is one of the most destructive paddy diseases, and it may cause grain loss of up to 60% -70%. The most favorable conditions are temperatures between 25-28 degrees, high humidity, high nitrogen levels, frequent rainfall, and a cloudy sky.

Brown Spot (Helminthosporium oryzae):

This disease is caused by a fungus that attacks the crop from the seedling stage to the milky stage. The disease symptoms will appear as minute spots on the coleoptile, leaf, and glume. The spots will be cylindrical, which later become circular and dark brown. Spots that appear on glumes cause grain discoloration, leading to seed germination failure and a reduction in grain quality and weight. While Excess nitrogen increases the severity of the disease, humidity over 80% and temperatures between 25-30 degrees are favorable conditions.

Sheath Rot:

This disease is also called Sarocladium oryzae. Its symptoms appear on the upper leaf. The disease will occur in irregular, greyish-brown spots, with an enlarged grey center and brown margins covering significant portions of the leaf. High nitrogen levels, close plantation, and temperatures between 25 and 30 degrees are favorable conditions. Through airborne and seed-borne conidia, the disease spreads.

Sheath Blight (Rhizoctonia solani):

From tillering to the heading stage, the fungus will affect the crop. The disease symptoms will appear on the leaf, typically near the water level. On the leaves, disease symptoms appear as elliptical, oval, or irregular greyish-grccn spots. The entire plant will result in death if the disease extends to the inner sheaths. Plants affected in the early stage result in poorly filled grain. Closer plantation, with 96% humidity, high nitrogen, and a temperature of 30-32 degrees, are the most favorable conditions.

Bacterial Diseases

Bacterial Leaf Blight (Xanthomonas oryzae):

In this disease, symptoms appear 1-2 weeks after transplantation, including Leaf Lesions, which are visible in seedlings. The disease can occur in earlier stages as well, but is mainly noticed during the heading stage. The increase in disease leads to the death of seedlings. Severe wind, application of heavy nitrogen during late top dressing, heavy rains, and temperatures between 25 °C and 30 °C are the most favorable conditions.

Bacterial Leaf Streak (Xanthomonas oryzae):

This disease manifests as narrow streaks of various lengths, characterized by a dark, greenish color. The lesions enlarge and affect more prominent veins, which turn from yellow-orange to brown. The disease spreads from one planting season to another through infected stubble and seeds. Severe wind, heavy rain, temperatures between 25-30 degrees, and deep irrigation water are the favorable conditions.

Viral Diseases

Rice Tungro Disease:

Rice tungro spherical or Rice tungro bacilliform virus. This disease can occur in both the nursery and the main field. The disease symptoms include plants being markedly stunted, and leaves discoloring from yellow to orange. The disease can affect the plant at all growth stages, particularly during the vegetative stage. The presence of vectors, virus sources, susceptibility, and the age of the host plant are favorable conditions.

Rice Grassy Stunt Disease:

Also known as Rice Grassy Stunt Tenuivirus. The symptoms of the disease include severely stunted plants due to excessive tillering and erect growth. Symptoms will develop after the plant becomes infected within 10-20 days and can survive until maturity, but panicles can't be produced. Growing rice around the year is the most favorable condition for spreading disease.

Rice Ragged stunt disease:

Also known as Rice Ragged stunt virus. The symptoms of the disease are ragged leaves with irregular margin formation and vein swelling. The disease causes delayed flowering and incomplete panicle emergence. The latent period of the disease is 3-35 days. The disease will not be transmitted congenitally. Continuous rice growing around the year is the most favorable condition for the spread of disease, as shown in Table **1**.

Table 1. Various paddy crop diseases, their symptoms, and favorable conditions.

Disease Name	Image of Diseased Plant	Symptoms	Favorable Conditions
Blast		Symptoms will appear on leaves, nodes, and glumes.	Temperature between 25-28 degrees, high humidity, high nitrogen, frequent rains, and cloudy sky.

(Table 1) cont.....

Disease Name	Image of Diseased Plant	Symptoms	Favorable Conditions
Brown Spot		Symptoms will appear as cylindrical spots, which later develop into circular areas with a dark brown color. Spots that appear on glumes cause grain discoloration, which leads to failure of seed germination, which in turn reduces grain quality and weight.	Excess nitrogen increases the severity of the disease. Humidity over 80% and temperature between 25-30 degrees.
Sheath Rot		Irregular greyish-brown spots and enlarged grey centers and brown margins cover significant portions of the leaf.	High nitrogen, close plantation, and temperature ranging between 25-30 degrees.
Sheath Blight		Elliptical, Oval, or irregular greyish-green spots on the leaf that are near the water level.	Closer plantation, humidity of 96%, high nitrogen, and temperature between 30-32 degrees.
Bacterial Leaf Blight		Kresek is the symptom seen in seedlings. The disease can occur in earlier stages as well, but is mainly noticed during the heading stage.	Severe wind, application of heavy nitrogen during late top dressing, heavy rains, and temperatures between 25-30 degrees.
Bacterial Leaf Streak		Narrow streaks of various lengths in dark-greenish color on veins. The lesions enlarge and affect more prominent veins, which turn from yellow-orange to brown.	Severe wind, heavy rain, temperature between 25-30 degrees, deep irrigation water.
Rice Tungro Disease		Plants will be markedly stunted, and leaves will discolor from yellow to orange.	Presence of vector, virus sources, susceptibility, and age of host plants.
Rice Grassy Stunt Disease		Plants will be severely stunted, exhibiting excessive tillering and an erect growth habit.	Continuous growing of the crop throughout the year.

(Table 1) cont.....

Disease Name	Image of Diseased Plant	Symptoms	Favorable Conditions
Rice Ragged Stunt Disease		Ragged leaves with irregular margins and vein swelling.	Continuous growing of the crop throughout the year.

LITERATURE SURVEY

T. Gayathri Devi *et al.* proposed "Image processing based rice plant leaves diseases in Thanjavur, Tamilnadu," which utilized 500 images from the Aduthurai research center in Thanjavur district [1]. The images were processed, and features were extracted and then classified using various classification methods, including ANN, KNN, multiclass SVM, and Bayesian classifiers. Images were preprocessed using a median filter and segmented for feature extraction using the K-Means clustering method. Image features were retrieved using DWT and SIFT methods, and the GLCM method was used for leaf texture analysis. This paper concluded that multiclass SVM is more accurate than other classifiers, with 98.63% accuracy.

"Rice Plant Leaf Disease Detection and Severity Estimation" was proposed by Radhika Wadhawan *et al.*, who worked on Rice diseases such as Blight, Brown Spot, and Leaf Smut [2]. This paper developed a Machine Learning model to scan images and identify infected leaves by comparing the area of infection [3]. This paper produced an accuracy of 85-86%.

Manoj Mukherjee *et al.* proposed "Damaged paddy leaf detection using image processing," which worked on Blast, Bacterial leaf blight, and Rice tungro paddy diseases [4]. This paper proposes a histogram-based concept for detecting damaged paddy leaves, where there is a noticeable difference in intensity between healthy and diseased paddy leaves. All the images used in this work were collected from various farms in the Burdwan district of West Bengal, India. The work produced classifier accuracy of 87% for Blast disease, 92% for Bacterial Leaf Blight disease, and 90% for Rice Tungro.

Eusebio L. Mique, Jr. *et al.*, proposed "Rice Pest and Disease Detection Using Convolutional Neural Network" [5]. This paper worked on the development of rice pests and disease detection applications. This paper utilizes transfer learning to train the model, which enables it to be trained faster than if it were trained from scratch by building a new model. The trained model achieved an accuracy of 90.9% with minimal cross-entropy.

Vimal K. Shrivastava *et al.* proposed "Rice Plant Disease Classification Using Transfer Learning of Deep Convolution Neural Network," which worked on Rice Blast, Bacterial Leaf Blight, and Sheath Blight diseases of the paddy crop [6]. This paper focuses on the automatic classification of rice plant diseases, utilizing transfer learning of deep CNNs to classify rice plant diseases [7]. In this work, various training-testing partitions were trained to achieve better accuracy. This paper achieved an accuracy of 91.37% with an 80%-20% training-testing partition.

Prabira Kumar Sethy *et al.* proposed "Image Processing Techniques for Diagnosing Rice Plant Disease: A Survey," where various papers from 2007 to 2018 were reviewed to help researchers learn about multiple paddy diseases in different growth stages [8]. This paper briefly discussed preprocessing, Segmentation, feature extraction, feature selection, and classification techniques for rice disease detection.

Pushpa Athisaya Sakila Rani *et al.* proposed "Paddy leaf symptom-based disease classification using deep CNN with ResNet-50" to discover rice leaf illness in rice plants. The input images were preprocessed using the median filter approach and then further segmented using the k-means clustering algorithm [9]. Using logistic regression, these segmented images were extracted and categorized using a deep CNN and ResNet-50 algorithm. This work achieved an accuracy of 97.3%.

Anandhan. K *et al.* proposed "Detection of Paddy Crops Diseases and Early Diagnosis Using Faster Regional Convolutional Neural Networks," which worked to detect and identify rice blast diseases [10]. This paper worked on Blast, Brown Spot, and Sheath Blight diseases [11 - 14]. This work achieved an accuracy of 82.5% for Faster R-CNN and 84.32% for Mask R-CNN during Training and 85.4% for Faster R-CNN and 87.45% for Mask R-CNN. This model can identify rice blast diseases, such as Blast (96%), Sheath Blight (94.5%), and Brown Spot (95%).

Ruoling Deng *et al.* proposed "Automatic Diagnosis of Rice Diseases Using Deep Learning," where six types of rice diseases were studied [15]. This model used five sub-models for training and testing: ResNet-50, ResNeXt-50, DenseNet-121, ResNeSt-50, and SE-ResNet-50. This work achieved an accuracy of 98% and an F1 score of over 0.95. This paper concluded that DenseNet-121, SE-ResNet-50, and ResNeSt-50 performed well among all the used sub-models [16].

Md Taimur Ahad *et al*. proposed a study titled "Comparison of CNN-based deep learning architectures for rice diseases classification," which examined various rice diseases and compared different CNN architectures [17]. In this paper, through transfer learning, various CNN architectures achieved an accuracy of 97% for DenseNet121, 96% for Seresnext101, 95% for EfficientNet, and 92% for Xception. The ensemble model in this paper achieved an accuracy of 97.62%.

ALGORITHMIC ANALYSIS

Various machine learning, deep learning algorithms, and image processing techniques were used for plant leaf disease detection research. According to the researchers, all the algorithms have both merits and demerits in the main field. Any model developed will undergo the following steps [18]:

- **Image acquisition and preprocessing:** Dataset loading, image resizing, color correction, and removal of noisy and inconsistent data [19].
- **Segmentation:** Converting images to binary format, identifying the boundaries of different photos, and clustering.
- **Feature Extraction:** Texture, Color, and Shape Analysis [20].
- **Classification:** Deep learning models and machine learning classifiers will be utilized to categorize various disease types.
- **Detection/Post-processing:** Refine segmented regions, identify connected components, and perform accurate Segmentation [21].

Image Acquisition and Preprocessing

This is the first step in image processing. This step involves Image Acquisition, Resizing, Color Correction, and Noise Reduction. Image acquisition is the process of using cameras or other imaging devices to capture high-quality images of plants that are exhibiting disease symptoms. Image resizing guarantees consistency in analysis, lowers computational cost, and resizes images to a consistent size [22]. Color correction enhances the visibility of symptoms in photographs, adjusts the color balance, and compensates for lighting changes. Noise reduction is used to reduce noise and enhance image clarity by applying filters such as bilateral, median, or Gaussian blur [23].

Segmentation

Segmentation is a method used in computer vision to support object detection and related tasks, dividing a digital image into distinct groupings of pixels or image segments. This step will take place after image preprocessing. This step involves Thresholding, Clustering, and Semantic Segmentation. Thresholding isolates diseased areas, distinguishes the plant from the backdrop, applies a threshold to the image, and converts it to a binary format. Clustering is used to identify areas of interest and group pixels with similar color or textural features using clustering methods such as K-means or mean shift clustering. Semantic Segmentation utilizes sophisticated methods to identify each pixel in the image as either healthy or diseased, enabling accurate symptom localization [24].

K-means Clustering

An unsupervised learning algorithm called K-Means Clustering separates the area of interest from the background. It separates the provided data into K clusters or portions based on K-centroids. This is used for unlabeled data. Finding particular groups based on data similarity, with K being the number of groups. K-means clustering aims to minimize the total squared distances between each point and the cluster center. Squared lengths connect each point to the cluster center [25].

Feature Extraction

The dimensionality reduction method, also known as feature extraction, is used to divide and reduce an initial group of raw data into more manageable subsets. The feature extraction technique is particularly useful for large datasets where there is a need to reduce the number of resources without sacrificing any significant or pertinent information. Through feature extraction, the amount of redundant information in the data collection will be reduced. The learning and generalization phases of the machine learning process will accelerate data reduction and facilitate the construction of the model with less computational effort. Feature extraction involves Texture Analysis, Color Analysis, and Shape Analysis. Texture Analysis is used to identify patterns associated with certain diseases and extract texture features using techniques such as Gabor filters, Local Binary Patterns, or Grey Level Co-occurrence Matrices. Color Analysis describes the color distribution of healthy and diseased areas, analyzing color properties such as color histograms and color moments. Shape Analysis is used to characterize the morphology of lesions or symptoms, extracting shape parameters such as area, perimeter, and compactness.

Image Classification

Image classification is crucial in computer vision due to its significance in contemporary technologies. It comprises labeling or tagging an entire image based on previously collected training data from already categorized images. This involves image analysis at the pixel level to identify the most suitable label for the image, providing valuable information for informed choices and feasible solutions. In this step, Machine learning classifiers are used to categorize image regions as healthy or diseased. Deep learning models are used to train and automatically extract characteristics from photos, classifying them into various illness categories.

Various machine learning and deep learning algorithms are used in plant disease detection.

Convolutional Neural Networks: By enabling automated and accurate identification of diseases from plant leaf images, CNN has revolutionized plant disease detection. In CNN, labeled images containing diseased and healthy plant leaves will be collected and resized to a standard size. To output class probabilities, a CNN stacks several convolutional layers followed by activation functions, pooling layers, fully connected layers, and a SoftMax activation function as the final layer. CNN trains the model using a labeled dataset where hierarchical features are extracted from images and mapped to relevant disease classes. The CNN model will be validated using a separate validation dataset to ensure the model is generalized for unknown data. After validation, the CNN model will be automatically deployed to real-time applications to diagnose plant diseases using farmers' captured images. Advantages of CNN in plant disease detection include Scalability, Robustness, and Feature learning.

Transfer Learning with CNNs: This is a powerful technique for improving plant disease detection performance using pre-trained models with limited labeled data. Transfer Learning utilizes a pre-trained CNN model that has already learned generic features, such as shapes, textures, and edges. To transform input images into high-dimensional feature vectors, the last layers of the pre-trained model will be removed by leaving the convolutional base intact. The pre-trained model will be fine-tuned on a small dataset of labeled plant disease images, with one fully connected layer added on top of the base to map the extracted features to the output classes. Train a modified CNN model using the plant disease dataset labeled Dataset. Data Augmentation techniques are used to prevent overfitting. Validate the trained model's performance on the validation dataset to assess its ability to generalize to unknown data. Evaluate model accuracy, recall, and other metrics to measure the efficiency of plant disease detection.

Support Vector Machines: Due to their ability to effectively handle high-dimensional data, SVMs are widely used in plant disease detection. This algorithm utilizes a labeled image dataset of plant images, where each image is associated with a corresponding label. The images are resized to a standard size by extracting relevant features that distinguish between healthy and diseased leaves. Each image in the Dataset is represented as a vector. The SVM classifier is trained with labeled training data. Evaluate the performance on separate validation datasets and measure metrics such as accuracy and precision to assess the classifier's ability to distinguish between healthy and diseased plant leaves. Fine-tune the hyperparameters and deploy the trained classifier to real-time applications, enabling the automatic detection of plant diseases from farmers' captured images.

K-Nearest Neighbors: This algorithm is simple and easy to understand, and it can be utilized for plant disease detection in this model. Plant leaf images labeled Dataset are used, where each image is linked to a label, and the image is resized to a standard size. Size and extract relevant features to determine healthy and diseased leaves. Each image in the Dataset will be represented as a feature vector that contains extracted features. There will be no external training in KNN, and the model will memorize the training data. In this model, the feature vector of the new image and the training images were calculated when the unlabeled image was present, along with the K nearest neighbors to the new image. Validate the trained model's performance on the validation dataset to assess its ability to generalize to unknown data. Evaluate model accuracy, recall, and other metrics to measure the classifier's performance, as in Table **2** [26].

Table 2. Advantages and disadvantages of various machine learning and deep learning models.

Machine Learning / Deep Learning Model	Advantages	Disadvantages
Convolutional Neural Networks	• Feature Learning • Spatial Hierarchical Representation • Generalization • Scalability	• Data Requirements • Computational Complexity • Interpretability • Overfitting • Domain Adaptation.
Support Vector Machines	• Effective in High-Dimensional Spaces • Robust to Overfitting • Nonlinear Classification • Optimal Margin Classifier • Versatility.	• Sensitivity to Hyperparameters • Computationally Intensive • Limited Interpretability • Scalability • Handling Imbalanced Data.
Random Forest	• Robustness to Overfitting • High Accuracy • Feature Importance • Handle Nonlinear Relationships • Scalability	• Lack of Interpretability • Memory and Computational Resources • Bias-Variance Tradeoff • Imbalance Handling. •
K-Nearest Neighbors	• Simplicity • Non-Parametric Adaptability • No Training Phase • Localized Decision Boundary	• Computational Complexity • Memory Usage • Curse of Dimensionality • Choice of Distance Metric • Imbalanced Data.

(Table 2) cont.....

Machine Learning / Deep Learning Model	Advantages	Disadvantages
Transfer Learning	• Feature Reuse • Reduced Training Time and Data Requirements • Improved Generalization • Domain Adaptation • Effective Feature Extraction	• Domain Mismatch • Limited Adaptability • Dependency on Pre-trained Models • Overfitting • Complexity and Hyperparameter Tuning.
Ensemble Model	• Improved Accuracy • Robustness to Overfitting • Model Stability • Versatility • Feature Combination	• Increased complexity • Computational Resources • Interpretability • Hyperparameter Tuning • Overfitting Risk
Decision Trees	• Interpretability • No Assumptions about Data Distribution • Feature Importance • Handling Mixed Data Types • Scalability	• Overfitting • High Variance • Bias towards Majority Classes • Limited Expressiveness • Model Fragility
Generative Adversarial Networks	• Data Augmentation • Realistic Image Generation • Unsupervised Learning • Domain Adaptation • Feature Learning	• Model Collapse • Training Instability • Quality Control • Generalization Performance • Data Privacy Concerns
Autoencoders	• Feature Learning • Dimensionality Reduction • Noise Reduction • Data Augmentation • Unsupervised Learning	• Limited Supervised Learning • Loss of Information • Complexity and Hyperparameter Tuning • Limited Diversity in Generated Samples • Dependency on Data Quality

Ensemble Methods: This is a powerful technique that combines multiple machine learning models to enhance accuracy and robustness in plant disease detection. Plant leaf images with labeled datasets are employed, and each image is linked to a labeled class that will be used in this model. The images are resized to the standard size, and relevant features are extracted. The selection of various models to form an ensemble takes place, where each model can be distinct. Each model is trained on a labeled Dataset and is prevented from overfitting using a cross-validation technique. In ensemble model construction, voting, weighted voting, bagging, boosting, and stacking are employed, where the predictions of all models are combined based on their performance. The trained model's performance is validated on the validation dataset to assess its ability to generalize to unknown data. The model accuracy, recall, and other metrics are evaluated to measure the ensemble's performance.

Detection/Post-Processing

After completing the initial processing steps, the final enhancement of images occurs in this step, where additional algorithms are applied to improve the quality and appearance of the images. Various post-processing techniques are:

- Noise Reduction using median filtering, Gaussian Filtering, and Wavelet denoising.
- Image enhancement using histogram equalization, contrast stretching, and unsharp masking.
- Edge Detection and Enhancement using the Sobel operator, Canny Edge detector, and Laplacian of Gaussian.
- Feature-level fusion, pixel-level fusion, and decision-level fusion will be used for Image fusion.

- Color correction using remote sensing, medical imaging, white balance, and color correction for color cast.
- Image registration utilizes matching corresponding features and applies geometric transformations to align images accurately.
- Image segmentation using morphological operations, region merging, and boundary refinement.
- Artifact Removal removes unwanted portions from images, thereby improving the quality and clarity of the images.

CONCLUSION

Various techniques and algorithms used for plant disease detection were discussed in this paper. Multiple works on Rice plant disease detection were evaluated. This paper highlights various machine learning and deep learning algorithms used for plant disease detection, along with their advantages and disadvantages. According to the carried research, each model has its advantages and disadvantages, so one must choose a research model based on the dataset size and other work requirements. The accuracy and functionality of the developed model depend on the selected machine learning or deep learning algorithm.

REFERENCES

[1] T. Gayathri Devi, and P. Neelamegam, "Image processing based rice plant leaves diseases in Thanjavur, Tamilnadu", *Cluster Comput.,* vol. 22, no. S6, pp. 13415-13428, 2019. [http://dx.doi.org/10.1007/s10586-018-1949-x]

[2] R. Wadhawan, M. Garg, and A. K. Sahani., "Rice plant leaf disease detection and severity estimation", *International Conference on Industrial and Information Systems(ICIIS), 15th IEEE,* 2020.

[3] "Rice: Index: Diseases of Rice." http://www.eagri.org/eagri50/PATH272/lecture01/index.html

[4] M. Mukherjee, T. Pal, and D. Samanta. "Damaged paddy leaf detection using image processing",

Journal of Global Research in Computer Science, Volume 3, No.10, October 2012.

[5] E.L. Mique Jr, and T.D. Palaoag, "Rice pest and disease detection using convolutional neural network", *Proceedings of the 1st International conference on Information Science and Systems,* pp. 147-151, 2018.
[http://dx.doi.org/10.1145/3209914.3209945]

[6] V.K. Shrivastava, M.K. Pradhan, S. Minz, and M.P. Thakur, "Rice plant disease classification using transfer learning of deep convolution neural network", *Int. Arch. Photogramm. Remote Sens. Spat. Inf. Sci,* vol. XLII-3, no. W6, pp. 631-635, 2019.
[http://dx.doi.org/10.5194/isprs-archives-XLII-3-W6-631-2019]

[7] "TNAU agritech oortal: Crop protection."
https://agritech.tnau.ac.in/crop_protection/rice_diseases/rice_8.html

[8] P. K. Sethy, N. K. Barpanda, A. K. Rath, and S. K. Behera. "Image Processing Techniques for Diagnosing Rice Plant Disease: A Survey", *International Conference on Computational Intelligence and Data Science (ICCIDS), 2019.*

[9] N. P. A. S. Rani and S. Singh. "Paddy leaf symptom-based disease classification using deep CNN with ResNet-50", *International Journal of Advanced Science Computing and Engineering,* vol. 4, no. 2, pp. 88-94, 2022.

[10] K. Anandhan and A. S. Singh. "Detection of Paddy Crops Diseases and Early Diagnosis Using Faster Regional Convolutional Neural Networks", *International Conference on Advance Computing and Innovative Technologies in Engineering (ICACITE),* Department of Electrical & Electronics Engineering, Galgotias College of Engineering and Technology, Gr. Noida, India, 2021.

[11] D. Gosai, B. Kaka, D. Garg, R. Patel, and A. Ganatra, "Plant Disease Detection and classification Using Machine Learning Algorithm", *International Conference for Advancement in Technology (ICONAT),* 2022.Goa, India
[http://dx.doi.org/10.1109/ICONAT53423.2022.9726036]

[12] Shima Ramesh, Mr. Ramachandra Hebbar, Niveditha M, Pooja R, Prasad Bhat N, Shashank N, Mr. P V Vinod, "Plant Disease Detection Using Machine Learning", *International Conference on Design Innovations for 3Cs Compute Communicate Control*, 2018.

[13] D.M, D.M, D.M, D.M, and D.M, "Paddy Diseases: causes, symptoms, and treatment options," Kisanvedika, BigHaat, Jun. 12, 2024. https://kisanvedika.bighaat.com/crop/major-diseases-of-paddy/

[14] X. Yang and T. Guo. "Machine Learning in plant disease research", *European Journal of BioMedical Research*, 2017.
[http://dx.doi.org/10.18088/ejbmr.3.1.2017.pp6-9]

[15] R. Deng, M. Tao, H. Xing, X. Yang, C. Liu, K. Liao, and L. Qi, "Automatic diagnosis of rice diseases using deep learning", *Front. Plant Sci.,* vol. 12, p. 701038, 2021.
[http://dx.doi.org/10.3389/fpls.2021.701038] [PMID: 34490004]

[16] E. Moupojou, A. Tafne, F. Retraint, A. Tadonkemwa, W. Dongmo, H. Tapamo, and M. Nkenlifack. "FieldPlant: A datasetof field plant images for plant disease detection and classification with deep learning", IEEE, 2023.

[17] M. T. Ahad, Y. Li, B. Song, and T. Bhuiyan. Touhid Bhuiyan, "Comparison of CNN-based deep learning architectures for rice diseases classification", *Artificial intelligence in agriculture*, 2023.
[http://dx.doi.org/10.1016/j.aiia.2023.07.001]

[18] W. Shafik, A. Tufail, A. Namoun, L. C. De Silva, and R. A. A. H. M. Apong. "A systematic literature review on plant disease detection: Motivations, classification Techniques, Datasets, Challenges, and Future Trends", *IEEE,* 2023.

[19] P. Sahu, A. P. Singh, A. Chug, and D. Singh. "A systematic literature review of machine learning techniques deployed in agriculture: A Case Study of Banana Crop", *IEEE*, 2022.

[20] Md. Mehedi Hasan, "M. M. Hasan, A. F. M. S. Uddin, M. R. Akhond, M. J. Uddin, M. A. Hossain, and M. A. Hossain. “Machine learning and image processing techniques for rice disease detection: A critical analysis”", *Int. J. Plant Biol.,* 2023.

[21] M. K. Roy, P. Manna, R. Roy, P. Roy, S. Rakshit, and M. K. H. Mondal. “Disease Detection in Paddy Crop using Machine Learning Techniques", *International Conference on Sustainable Computing and Data Communication Systems (ICSCDS),* 2023.

[22] K. Sonar, A. Angaitkar, S. Sarode, P. Harshe, S. Chavhan, and S. N. Wasankar. "Plant disease identification using SVM", *International Journal ofCreative Research Thoughts (IJCRT),* Volume 10, Issue 3, March 2022.

[23] M. A. Hussein and A. H. Abbas. "Plant leaf disease detection using support vector machine", Al-Mustansiriyah Journal of Science, Volume 30, Issue 1, 2019.

[24] S. D. Khirade, A.B. Patil, "Plant disease detection using image processing", *International conference on Computing Communication Control and Automation,* 2015.

[25] J. Liu and X. Wang. "Plant diseases and pests detection based on deep learning: a review", plant methods 17, Article Number 22, 2021.

[26] B. Tugrul, E. Elfatimi, and R. Eryigit. "Convolutional Neural Networks in Detection of Plant Leaf Diseases: A Review", *Advances in Agricultural Engineering Technologies and Application,* 12(8), 2022.

CHAPTER 2

Review on an Intelligent System for Quality Evaluation of Dry Fruits Using Hyperspectral Imaging

Shweta Bhelonde[1,*] and **Manoj B. Chandak**[1]

[1] *Department of Computer Science and Engineering, Ramdeo Baba University, Nagpur, Maharashtra, India*

Abstract: In the rapidly evolving field of agricultural technology, the need for advanced, reliable methods for assessing dry fruit quality has become paramount. This paper presents an empirical review of the rapidly increasing application of Machine Learning (ML) and Deep Learning (DL) techniques in hyperspectral imaging for the qualitative analysis of dry fruits and their subtypes. Traditional methodologies, while having laid a robust foundation, often grapple with constraints such as limited precision, suboptimal accuracy, and scalability challenges. Furthermore, these conventional approaches typically exhibit significant delays in processing and struggle with complexity in handling diverse dry fruit categories. Addressing these limitations, this review comprehensively evaluates ML and DL methodologies tailored explicitly for hyperspectral imaging applications in dry fruit quality analysis. The proposed review process stands out by precisely comparing these methods across a spectrum of critical evaluation metrics, such as precision, accuracy, recall, processing delay, complexity, and scalability. This approach not only bridges the gaps identified in existing literature but also lays the groundwork for a more nuanced understanding of the capabilities and limits of these advanced technologies in practical scenarios. This, in turn, has improved quality control measures and enhanced overall productivity levels. In conclusion, this work extends the existing body of knowledge. It sets a new benchmark for applying deep learning techniques in agricultural hyperspectral imaging, thus marking a significant stride forward in pursuing agricultural innovation and excellence.

Keywords: Deep learning, Dry fruit quality, Hyperspectral imaging, Machine learning, Scenarios.

* **Corresponding author Shweta Bhelonde:** Department of Computer Science and Engineering, Ramdeo Baba University, Nagpur, Maharashtra, India; E-mail: shwetabhelonde@gmail.com

Suneeta Satpathy, Sachi Nandan Mohanty & Subhendu Kumar Pani (Eds.)

INTRODUCTION

The introduction of advanced computational techniques, particularly Machine Learning (ML), Deep Learning (DL), and Generative AI [1 - 3] into the domain of agricultural technology has foreshadowed a new era in dry fruit quality assessment. This paper delves into the intersection of these cutting-edge technologies with hyperspectral imaging, a method gaining traction due to its non-destructive nature and efficacy in the quality evaluation of dry fruits. The essence of this study lies in its meticulous review and analysis of existing ML and DL methodologies applied in hyperspectral imaging for dry fruit quality assessment, highlighting their capabilities and demystifying their limitations.

Historically, dry fruit quality assessment has relied heavily on manual inspection and conventional imaging techniques. While fundamental in their contributions, these methods [4 - 6] often falter in precision, scalability, and efficiency, mainly when dealing with large-scale agricultural operations. In contrast, hyperspectral imaging emerges as a powerful alternative, offering detailed insights into the internal quality of dry fruits without physical intrusion. Integrating ML and DL with hyperspectral imaging has amplified its potential, enabling more nuanced, accurate, and rapid assessments.

However, this integration is not without challenges. The complexity inherent in ML and DL algorithms, the diverse nature of dry fruits, and their varying hyperspectral properties pose significant hurdles. Additionally, the vast array of available ML and DL models, each with distinct strengths and weaknesses, necessitates a thorough and comparative analysis to determine the most effective approaches for specific dry fruit quality assessments.

Early dry fruit quality assessment studies predominantly focused on traditional methods, manual inspection, and basic imaging techniques. Though effective to a certain extent, these methods were limited by their subjectivity, lack of scalability, and inability to detect internal defects. The advent of hyperspectral imaging marked a significant leap forward. It provides a non-invasive way to assess dry fruit quality, leveraging the temperature variations and hyperspectral properties of dry fruits. However, interpreting hyperspectral images was initially challenging, often requiring extensive expertise and subjectivity in analysis.

The integration of ML into this domain [10 - 12] marked another pivotal moment. Early research in this area demonstrated that ML algorithms could effectively classify dry fruits based on quality by analyzing patterns in hyperspectral images. These studies laid the groundwork for more sophisticated applications, highlighting the potential of ML in automating and enhancing the accuracy of dry fruit quality assessment.

Subsequent research delved into applying DL, particularly Convolutional Neural Networks (CNNs), in interpreting hyperspectral images. This strand of research revolutionized the field, offering unprecedented accuracy in detecting external and internal defects in dry fruits. The ability of DL models to learn from vast amounts of data and identify intricate patterns undetectable to the human eye was a game-changer.

However, the literature also reveals specific gaps and challenges. One recurring theme is the complexity and resource intensity of DL models, which pose challenges in computational requirements and practical applicability in resource-limited settings. Another area highlighted in the literature is the need for more extensive datasets encompassing a wider variety of dry fruits and conditions to enhance the robustness and generalizability of these models. By systematically evaluating various ML and DL models in the context of hyperspectral imaging for dry fruit quality assessment, this paper aims to provide a clear, structured understanding of the current landscape while identifying areas ripe for future research and development processes.

MOTIVATION AND CONTRIBUTIONS

The Motivation and Contribution section of this paper centers on a central thesis: the integration of Machine Learning (ML), Deep Learning (DL), and Generative AI with hyperspectral imaging in agricultural technology is not merely a progressive step but a necessary evolution in the pursuit of optimal dry fruit quality assessments. This conviction is rooted in a confluence of factors, primarily the escalating demand for high-quality agricultural products and the relentless pursuit of technological innovation in the agriculture scenarios.

For instance, from Fig. (**1**), hyperspectral imaging enables the distinction between good-quality figs and bruised figs by capturing subtle spectral differences that conventional methods often overlook. Good-quality figs display consistent spectral profiles, indicating uniformity in their texture and chemical composition. In contrast, bruised figs exhibit disrupted spectral patterns due to internal damage, moisture variation, or discoloration, manifesting as identifiable anomalies in hyperspectral data. When coupled with advanced ML and DL techniques, these insights allow for highly accurate, non-invasive quality assessments, ensuring that only the best produce reaches consumers.

The motivation for this study emanates from several critical observations. First, the burgeoning global population and the accompanying food demand underscore the need for efficient, scalable, and accurate methods for assessing agriculture quality. Traditional methods, though foundational, are increasingly inadequate in meeting these demands, especially in the context of quality assessment of dry fruits. Hyperspectral imaging, in this regard, offers a compelling alternative, given its non-invasive nature and the depth of information it can provide. However, the true potential of hyperspectral imaging in this domain remains largely untapped, mainly, a gap this paper seeks to address.

Fig. (1). Hyperspectral images for good quality fig *vs.* bruise fig [1].

Further motivation is derived from the dynamic nature of Machine Learning (ML) and Deep Learning (DL) technologies. These computational methods have demonstrated remarkable success in various fields, yet their application in agricultural hyperspectral imaging is still nascent. This paper, therefore, endeavors to explore and elucidate the efficacy of these technologies in enhancing the accuracy and efficiency of dry fruit quality assessment, thereby contributing to the broader field of precision agriculture scenarios.

The contributions of this study are multifaceted. Firstly, it offers a comprehensive review of current ML and DL methods applied to hyperspectral imaging in dry fruit quality assessment, filling a significant gap in the existing literature. Secondly, by comparing these methods across various evaluation metrics, such as precision, accuracy, recall, processing delay, complexity, and scalability, the paper provides valuable insights into the strengths and weaknesses of each approach. This comparative analysis is not just theoretical; it has practical implications, guiding practitioners in selecting the most suitable methods for their specific needs.

Moreover, this study contributes to the field by highlighting the challenges and opportunities in integrating ML and DL with hyperspectral imaging for dry fruit quality assessments. It identifies key areas where further research and development are necessary, thus paving the way for future innovations in this field for different scenarios. In essence, the Motivation and Contribution section establishes the significance of this research within the broader context of agricultural technology sets. It underscores the urgency for advanced dry fruit quality assessment methods, driven by global food security concerns and the quest for agrarian efficiency levels. This paper's contribution lies in its comprehensive review of existing methodologies, critical analysis, and suggestions for future advancements. By charting a course for further research and practical application, this study serves as a beacon for those navigating the complex intersection of machine learning, deep learning, and hyperspectral imaging to pursue agricultural excellence. Ultimately, it aims to catalyze a shift toward more sophisticated, reliable, and scalable dry fruit quality assessment methods, significantly impacting precision agriculture and beyond for different scenarios.

REVIEW OF EXISTING MODELS

Researchers propose various models for the hyperspectral analysis of dry fruit image samples. From Table **1**, for instance, work by A. Kamaruzaman (2022) proposes a deep learning methodology for detecting and localizing fig fruits in complex environmental pictures, achieving high accuracy with the YOLOv4 model process. Work in (2022) A. Vinci introduces UAV technology for canopy characterization in hazelnut orchards, offering a more efficient and accurate method than manual measurements [4]. Work by A. Zeb (2022) focuses on the potential of short-wave NIRS for fruit classification, primarily based on the fruit's O-H and C-H overtone features. Various machine learning models were trained to classify fruits using spectral feature vectors, achieving high accuracy in fruit classification based on the absorptivity of shortwave NIR radiation sets. D. Buonocore discusses the development of an intelligent sensing system for monitoring the fig drying process, validating its effectiveness in speeding up the drying process [8]. The author, E. Kizilay, investigates the use of microwave properties to assess the freshness of walnuts. The study simulates different walnut samples and analyzes Transmission Coefficients (TCs) from pyramidal horn antennas. The results show a significant difference in TC for fresh walnuts, allowing for the assessment of walnut freshness [13].

Table 1. Empirical evaluation of reviewed models.

Author Name	Method Used	Findings	Results	Limitations
A. Kamaruzaman	Deep learning (YOLOv3, YOLOv4).	Fast and effective detection of fig fruits using deep learning.	YOLOv4 outperformed YOLOv3 with higher average precision (90.02%).	Limited fig fruit detection in images.
A. Vinci	UAV technology.	Automated hazelnut canopy characterization using UAVs.	Good performance in canopy width and volume evaluation. Height measurement had limitations.	Limited to hazelnut canopy characterization.

(Table 1) cont.....

Author Name	Method Used	Findings	Results	Limitations
A. Zeb	Short-wave NIRS, QDA classifier.	Short-wave NIRS is used for fruit classification. QDA achieved high accuracy.	Cross-validation accuracy of 100% and test data accuracy of 93.02% for 83 spectral features. Cross-validation accuracy of and test data accuracy of 90.38% for four features.	Dependency on short-wave NIR radiation.
Buonocore	Intelligent sensing system for the fig-drying process.	Monitoring the fig drying process using an innovative sensing system.	Improved fig drying process monitoring with temperature and humidity data samples.	Limited to the fig drying process monitoring.
E. Kizilay	Microwave transmission coefficients.	Simulated walnut freshness assessment using microwave properties.	Significant differences in Transmission Coefficient (TC) for fresh walnuts. RMSE was used for comparison.	Limited to walnut freshness assessment.
J. Wei	Machine learning (Random Forest, Extra Tree, Decision Tree).	Classification of raisin types using different models.	Random Forest had the highest recognition success rate. Primary Axis Length is the decisive feature.	Limited to classification of raisin types.
L. Zabawa *et al.*,	Fully convolutional neural networks.	Detection of individual berries in grapevine images.	Accurate detection of white grapes in vineyards (accuracy 94.0% for VSP, 85.6% for SMPH).	Limited to single-berry detection in vineyard images.
M. Dejanov,	VIS and NIR spectroscopy, PLSR, SVMR.	VIS-NIR and NIR used for β-carotene prediction in apricots. PLSR and SVMR models with good results.	R^2c = 0.91, RMSEC = 20.1 (calibration), R^2v = 0.84, RMSEV = 27.1(validation) for PLSR in the VIS-NIR range. R^2c = 0.92, RMSEC = 19.2 (calibration), R^2v = 0.84, RMSEV = 26.7 (validation) for SVMR.	Poor performance with the second derivative Savitzky-Golay. Large variance in the drying process model.
M. Greco *et al.*	Transmission mode Setup Organizing Map (SOM).	Distinguishing healthy and rotten hazelnuts using transmission mode setup.	SOM with Kohonen algorithm used for hazelnut quality classification.	Limited to hazelnut quality classification.

(Table 1) cont.....

Author Name	Method Used	Findings	Results	Limitations
M. Greco	97 GHz imaging system.	Discrimination of healthy and rotten hazelnuts using THz radiation.	THz technology as a non-invasive method for hazelnut quality control.	Limited to hazelnut quality discrimination.
N. E. Md Rosli	Image segmentation (Threshold, K-means, Sharp U-Net).	Image segmentation for fig fruit separation from the background.	Sharp U-Net demonstrated the highest accuracy in fig image segmentation.	Limited to fig fruit image segmentation.
O. Sahin	Machine learning, feature selection with Random Forest.	Classification of raisin species using morphological features.	Multi-Layer Perceptron achieved 94.07% accuracy using the most essential feature. Outperformed previous models.	Limited to raisin species classification.
R. Wang, N. Tuerxun and J.Zheng	Multi-spectral UAV images, BP neural network.	Impact of scale effects on CLC estimation using UAV images. BP neural network model.	Best estimation accuracy at 120m flight altitude (R^2 = 0.69, RMSE = 2.67μg/cm^2, NRMSE = 5.47%)., Boruta and BP neural networks are the best models.	Limited to specific flight altitudes.
S. Bayrakdar	Image processing for hazelnut type and quality.	Classification of shelled hazelnuts based on size and shape characteristics.	Achieved 84% accuracy in grouping hazelnuts by type and quality.	Limited to shelled hazelnut classification.
Y. Tang	Spatial Attention Grape Berry Counting Network (SAGBCNet).	Automated grape berry counting in vineyards using deep learning.	SAGBCNet achieved lower counting errors in grape berry counting—high accuracy (MAE 35.39, MSE 47.39).	Limited to grape berry counting in vineyards.

The author, J. Wei [14 - 16], utilizes three machine learning models to distinguish types of raisins. Random Forest emerges as the most successful model, with Major Axis Length being the decisive feature for classifications [17 - 19]. The work (2019) by L. Zabawa *et al.* [20] applies fully convolutional neural networks to count single berries in grapevines, offering a non-invasive and accurate alternative to manual analysis [21 - 23]. The author M. Dejanov [24 - 26] investigated the use of Visible and Near Infrared spectroscopy (VIS and NIR) to predict β-carotene content in apricots. Two regions, VIS-NIR and NIR, were analyzed for prediction ability using Partial Least Squares Regression (PLSR) and Support Vector Machine Regression (SVMR) process. The results indicate that the VIS-NIR range yields the best model performance for predicting β-carotene contents of various dry fruits. M. Greco *et al.* describe an experimental setup using transmission mode in the 18-40 GHz frequency range to distinguish healthy and rotten hazelnuts, employing a Self-Organizing Map (SOM) for classifications [27 - 29]. Work in (2022) M. Greco explores the use of THz radiation for discriminating between healthy and rotten hazelnuts. The study employs a 97 GHz imaging system for such purposes.

Work in (2022) N. E. Md Rosli [30] focuses on fig-fruit image segmentation, comparing different methods and finding Sharp U-Net to be the most effective for the segmentation process [31] and scenarios. O. Sahin (2023) [32] presents machine learning models for classifying different raisin species based on morphological features. Feature selection using Random Forest determines the most critical features, achieving a high accuracy rate in the raisin classification process [33]. In 2023, R. Wang, N. Tuerxun, and J. Zheng's study reconnoiter the impact of scale effects on the accuracy of Chlorophyll Content (CLC) estimation in vegetation using multi-spectral UAV images at different flight altitudes. The results suggest that different flight altitudes significantly affect CLC estimation, with the best accuracy achieved at 120 m flight altitude levels [34 - 36]. Y. Tang (2022) presents a Spatial Attention Grape Berry Counting Network (SAGBCNet) for estimating grape yields, achieving lower counting errors and high accuracy levels [37 - 39].

An Empirical Evaluation of Reviewed Models can be discussed in Table **1** as follows:

From Table **2**, as per B. Yalçın (2015), hazelnut orchards are economically vital in Turkey, and remote sensing images are essential for their detection and management. This study introduces a Selective Sampling-based Approximate Spectral Clustering Ensemble (SSASCE) for extracting hazelnut fields, addressing computational challenges and demonstrating its superiority levels [40 - 42]. As per D. Di Caro (2020) [43 - 45], TD-NMR is a valuable tool for monitoring food quality control levels. This article proposes an inline quality evaluation method for in-shell hazelnuts, focusing on signal processing to reduce execution time and classify defective nuts from good ones. According to D. Di Caro (2018), Non-destructive testing is essential for product quality assessment. TD-NMR is utilized for evaluating in-shell hazelnuts, focusing on the amplitude of the Free Induction Decay (FID) signal and CPMG signal components, offering a rapid inline quality evaluation method and process. As per D. Di Caro (2017), the flavor and taste of hazelnuts are essential in the confectionery industry. This paper utilizes Time Domain - NMR (TD-NMR) to detect hidden defects in hazelnuts in their shells. The Carr-Purcell-Meiboom-Gill (CPMG) technique is employed to evaluate transverse relaxation Times (T2) for healthy and unhealthy hazelnuts, offering a basis for the classification process. As per H. Kalkan (2008), Aflatoxin-producing molds can affect the quality of hazelnuts. This research employs multispectral imaging to separate contaminated hazelnut kernels from healthy ones, highlighting the importance of specific spectral bands for detecting aflatoxin contamination.

Table 2. Empirical evaluation of machine learning methods.

Author Name	Method Used	Findings	Results	Limitations
B. Yalçın	Approximate Spectral Clustering Ensemble method (SSASCE).	Introduced Selective Sampling, approximate Spectral Clustering Ensemble.	Outperformed spectral clustering for hazelnut field extraction.	Dependent on Dataset and application.
D. Di Caro	Inline quality evaluation using TD-NMR analysis.	Quality evaluation of in-shell hazelnuts based on TD-NMR analysis.	Developed an inline quality evaluation method for in-shell hazelnuts using TD-NMR analysis.	Focused on in-shell hazelnut quality evaluation.
D. Di Caro	TD-NMR analysis.	Quality evaluation of in-shell hazelnuts based on TD-NMR analysis.	Developed a method for inline quality evaluation of in-shell hazelnuts using TD-NMR.	Focused on in-shell hazelnut quality evaluation.
D. Di Caro	Time Domain NMR (TD-NMR).	Detection of hidden defects in hazelnuts in a shell based on TD-NMR.	Evaluated transverse relaxation times (T2) for healthy and unhealthy hazelnuts.	Focused on detecting hidden defects.
H. Kalkan	Multispectral imaging.	Separation of contaminated hazelnut kernels from healthy ones.	Identified discriminative bands for aflatoxin contamination detection.	Focused on separating contaminated hazelnuts.
I. Khosa	Automatic detection using X-ray images.	Real-time automatic detection, segmentation, and classification of hazelnuts using X-ray images.	Proposed a method for real-time automatic detection, segmentation, and classification of hazelnuts using X-ray images.	Focused on hazelnut detection, not classification.
I. Onaran	Acoustic signal analysis	Prototype system to detect empty hazelnuts using acoustic signals.	Classified 98% of fully developed kernels and 97% of empty kernels.	Limited to detecting empty hazelnuts.
I. Onaran	Sparse signal representation approach.	Separation of empty and full hazelnuts using impact acoustic signals.	Achieved Classification accuracies of 98.3% and 96.8% for undeveloped hazelnut separation.	Limited to separating empty and full hazelnuts.

(Table 2) cont.....

Author Name	Method Used	Findings	Results	Limitations
K. Taşdemir	Rule-based classification using remote sensing imagery.	Proposed a rule-based classification using multi-scale Gabor features and spectral values for an accurate hazelnut orchard.	Compared classification accuracies using WorldView-2 and Quickbird imagery for hazel nut orchard detection.	Limited to hazelnut orchard detection.
K.Taşdemir	Spectral and spatial information.	Detection of hazelnut orchards from VHR images.	Proposed a Self-Organizing Map (SOM) method for hazelnut orchard detection from QuickBird imagery.	Focused on a hazelnut orchard.
M. Demir, M. Ulutaş and H. Türe	Fuzzy Logic-based image processing.	Automated identification of partially skinned hazelnuts on a conveyor belt.	Improved classification accuracy for processed hazelnuts on a conveyor belt.	Limited to identifying partially skinned hazelnuts.
N. Kumar Naik	Quality assessment based on chips and scratches.	Automated almond quality evaluation based on chips and scratches.	Proposed a computerized method for assessing the quality of almonds.	Focused on quality assessment of almonds.
Ş. Aydın,	Open Data for hazelnut data management.	Investigated open data in the context of hazelnut data management.	Proposed a data processing model for agricultural open data in the context of hazelnuts.	Focused on data management, not hazelnut analysis.
S. A. Guvenc	Computer vision approach.	Classification of partly skinned, skin-removed, and rotten hazelnut kernels.	Achieved a classification accuracy rate of 93.57% for processed hazelnut kernels.	Limited to classifying hazelnut kernels.
Y. H. Habiboğlu	Impact vibration analysis.	Separation of empty and whole hazelnuts using impact vibration analysis.	Achieved 96.7% and 98.3% classification accuracies for Antep-style Turkish pistachio nuts and hazelnuts.	Limited to separating empty and whole hazelnuts.

As per I Khosa (2014), an automated method for real-time detection, segmentation, and classification techniques used for hazelnuts using X-ray images is presented for different use cases. This approach uses moment invariants, texture properties, and Principal Component Analysis (PCA) for classification, achieving a high correct classification rate of 98.6% for negative examples. As per I. Onaran (2005), detecting empty hazelnuts is crucial for quality and safety. This study uses vibration analysis to separate shell pieces from hazelnut and pistachio kernels. Various feature parameters and LSF values are extracted from vibration signals, achieving classification accuracies of 98% for fully developed kernels and 97%

for empty kernels. As I. Onaran (2007) states that sparse signal representation impacts acoustic signals to discriminate between empty and whole hazelnuts. This method achieves high classification accuracies of 98.3% and 96.8% for Antep-style Turkish pistachio nuts and hazelnuts. As per K. Taşdemir, remote sensing imagery is a valuable tool for agricultural management, especially when detecting permanent crops like hazelnuts. This study presents a rule-based classification approach using multi-scale Gabor features and spectral values to accurately detect hazelnut orchards, comparing results from WorldView-2 and Quickbird imagery sets. As per K. Taşdemir (2010), accurately detecting nut orchards from high-resolution (VHR) images is challenging due to the varying statistics. This paper presents a self-organizing map that combines spectral and spatial information to improve hazelnut orchard detection from Quickbird imagery sets.

According to M. Demir, M. Ulutaş, and H. Türe (2012), hazelnuts are a vital agricultural product in Turkey, the world's largest producer of hazelnuts. The hazelnut kernel is a key ingredient for making chocolate in the food processing industry. Partially skinned hazelnuts are rejected by chocolate manufacturers. This paper proposes an automated image processing and classification system based on Fuzzy Logic to identify partially skinned hazelnuts on a conveyor belt process. As per N. Kumar Naik (2023), almonds are renowned for their health benefits, and their quality assessment is crucial for consumers. This paper introduces an automated method for assessing almond quality, demonstrating its effectiveness through screenshots. As per Ş. Aydın (2018), Turkey plays a significant role in global hazelnut production and exportation, making data availability crucial for sustainable agricultural activities. This paper explores the "Open Data" concept in Turkish agriculture and proposes a data processing model for hazelnut-related data management processes [37]. As per S. A. Guvenc (2015), partially skin-removed hazelnut kernels, skin-removed kernels, and rotten hazelnut kernels pose challenges in the chocolate and biscuit industries in different scenarios. Manual selection methods are time-consuming and prone to error in real-time use cases for dry fruit types. This study classifies processed hazelnut kernels using a computer vision approach, achieving a classification accuracy rate of 93.57% [38].

As per Y. H. Habiboğlu (2013), impact vibration analysis is used to separate nut-shell pieces from hazelnut and pistachio kernels. Feature parameters and LSF values are extracted from vibration signals, achieving high classification rates for both nut types. An Empirical Evaluation of Machine Learning Methods can be discussed in Table **2** as follows:

In Table **3**, A. Bakiya (2023) proposed that analyzing leaf nutrition content is crucial for plant growth and fruit quality. A. C. Cruz (2022), Almond Leaf Scorch

Disease (ALSD) has emerged as a significant threat to almond production. Traditional disease diagnosis methods often require expert knowledge and third-party laboratory testing. However, recent advancements have harnessed the power of deep learning algorithms to detect ALSD through smart camera-captured images. These algorithms outperform human capabilities in terms of accuracy and speed. For instance, DenseNet201 achieved an accuracy of 88.72% in identifying ALSD levels. This breakthrough promises affordable and swift disease detection in almond orchards. Effective water management in agriculture is crucial, especially in semiarid regions. Agricultural and forest residues can be valorized to preserve ecosystems and produce derivative products. B. Mamadou Seydou (2022) investigates the thermal decomposition kinetics of co-products, including peanut shells, cashew nut shells, palm nut shells, and millet stems. It employs thermogravimetric analysis and kinetic modeling to determine the activation energy for different co-products. This research contributes to the development of sustainable agricultural practices. The study uses hyperspectral imaging to explore a rapid, nondestructive method for detecting and classifying moisture content in dried wolfberries. Divyanth, L.G., Chakraborty, S., Li, B. *et al.* (2022) tested various regression algorithms, showcasing the potential of hyperspectral imaging technology in ensuring the quality of agricultural products. Quality control in the tobacco industry requires accurate prediction of nicotine content.

Table 3. Empirical evaluation of different deep learning methods.

Author Name	Method Used	Findings	Results	Limitations
A. Bakiya	Hyperspectral Imaging for Leaf Nutrition.	Identification of leaf nutrition content.	SG-PLS accurately predicted leaf nutrition content for two different leaf types.	Limited to leaf nutrition content identification may not apply to other plant characteristics.
A. C. Cruz	Deep Learning with CNNs.	Detected Almond Leaf Scorch Disease (ALSD).	Achieved 88.72% accuracy in ALSD detection using DenseNet 201.	Limited to image-based detection, future work is needed for *in vivo* detection.
Divyanth, L.G., Chakraborty, S., Li, B. *et al.*	Spatial-Spectral Two-Branch-CNN Model.	Semantic segmentation of hyperspectral data.	Improved segmentation results in hyperspectral data, especially at class borders and in mixed pixels.	Focused on hyperspectral data segmentation may not be directly applicable to all agricultural tasks.
H. R. Seireg	Ensemble Machine Learning Algorithms.	Wild blueberry yield prediction.	Stacking Regression (SR) showed high performance with R^2 of 0.984 and RMSE of 179.898.	Limited to wild blueberry yield prediction and may not apply to other crops.

(Table 3) cont.....

Author Name	Method Used	Findings	Results	Limitations
J. M. Sánchez	METRIC and STSEB Energy Balance Model.	Monitoring actual evapotranspiration in orchards.	ETa results showed an average error of < 1.0 mm/day for METRIC and STSEB models using Landsat 8 images.	Limited to specific geographic regions (southeastern Spain) and satellite imagery data samples.
K. Susuki	Distillation Column Design.	Production of blueberry-derived alcoholic beverage.	Developed a mechanical design for distillation column achieving a production ratio of 1:7.	Focused on blueberry-derived average production, not applicable to other fruits or crops.
M. Yi, C. Zhao, F. Liao, and W. Yao	Improved Efficient Net Model.	Blueberry variety classification.	Achieved high accuracy in blueberry variety classification using an improved Efficient Net model.	Focused on blueberry variety classification, this approach may not be applicable to other crops.
Nirere, A	Hyperspectral Imaging for Pesticide Residues.	Analyzing pesticide residues in vegetables and fruits.	Analyzed the present situation of pesticide residues and proposed countermeasures.	Limited to pesticide residue analysis may not address other agricultural challenges.
P. B. Cubas Muñoz	YOLOv8 and Custom CNN Model.	Detection and categorization of blueberries.	A notable accuracy of 91.14% in blueberry classification based on ripeness levels.	Focused on blueberry classification, this approach may not be applicable to other fruits or vegetables.
P. Brayam Cubas Muñoz	Mask R-CNN and CNN Model.	Detection and Classification of blueberries.	Achieved a 90.74% accuracy in blueberry classification-based on ripeness levels.	Focused on blueberry classification, this approach may not be applicable to other fruits or vegetables.
T. R. Bayala, S. Malo and Z.Kinda,	DL4 Cashew Deep Learning Model.	Identification and counting of cashew trees.	Achieved 93% accuracy in cashew tree identification.	Limited to cashew tree detection and may not apply to other tree species.
U. Pestel-Schiller, Y. Yang and J. Ostermann	Hyperspectral Imaging for Nicotine Content.	Prediction of nicotine content in powdered tobacco.	PLSR outperformed other models in predicting nicotine content using hyperspectral data samples.	Specific to the tobacco industry, it may not be applicable to other crops or substances.

(Table 3) cont.....

Author Name	Method Used	Findings	Results	Limitations
Y. Chao and S. Yan	Hyperspectral Imaging for Leaf Nutrition.	Identification of Leaf Nutrition Content.	SG-PLS achieved an accurate prediction of leaf nutrition content using hyperspectral imaging.	Limited to leaf nutrition content identification may not apply to other plant characteristics.
Y. Wang, L. Suarez, D. Ryu and P. J. Zarco-Tejada	Solar Induced Chlorophyll Fluorescence.	Improved leaf N estimation with SIF and Cab.	Cx ranks third after Cab and SIF for explaining leaf N variability over two growing seasons.	The study focuses on leaf-level measurements, which may not apply directly to large-scale agriculture.

H. R. Seireg (2022) introduced ensemble machine learning algorithms to forecast wild blueberry yields. Stacking and cascading regression approaches result in impressive forecast accuracy levels. The study showcases the potential of ensemble machine learning for improving yield predictions.

The study by J. M. Sánchez, J. González-Piqueras, *et al.* evaluates the performance of two classic energy balance models, METRIC and STSEB, in monitoring actual Evapotranspiration (ETa) using Landsat 8 images and samples. This research demonstrates that these models can accurately estimate ETa, aiding in the detection of water stress at the field scale.

Utilizing blueberries to produce value-added products is an intriguing prospect. K. Suzuki, A. Ipanaque, and L. Vinces (2023) discuss the design of a distillation column for producing a blueberry-derived alcoholic beverage with antioxidant properties. Blueberry variety recognition is essential for ensuring quality control and effective inventory management processes. Two studies by M. Yi, C. Zhao, F. Liao, and W. Yao (2022) and P. B. Cubas Muñoz propose innovative methods for classifying blueberry varieties. The former introduces an improved EfficientNet model, achieving high accuracy in recognizing blueberry varieties across different stages. The latter combines YOLOv8 and a custom CNN to detect and classify blueberries based on ripeness levels, demonstrating notable accuracy.

Nirere, A. (2023) discusses the design of a distillation column for producing a blueberry-derived alcoholic beverage with antioxidant properties. Mathematical calculations and mechanical design considerations are presented to optimize the distillation process. This research offers insights into innovative uses of agricultural products. Maintaining the quality of dried wolfberries is essential for long-term utilization.

In 2023, T. R. Bayala introduced DL4 Cashew—a deep learning-based model—to identify cashew trees in heterogeneous agricultural areas. The model achieved a remarkable 93% performance in the tree identification process. The paper presents an algorithm for counting cashew trees, offering valuable insights into crop monitoring sets.

U. Pestel-Schiller (2022) suggested that Hyperspectral images are combined with chemometric models to predict nicotine levels in powdered tobacco samples. Partial Least Squares Regression (PLSR) outperforms other models, demonstrating the efficiency of hyperspectral imaging for quality control in the tobacco industry.

In 2022, Y. Chao and S. Yan proposed that hyperspectral imaging is employed to determine the nutrition content of leaves under different conditions. Regression algorithms accurately predict leaf nutrition content, including Savitzky-Golay Partial Least Squares Regression (SG-PLS). This research highlights the potential of hyperspectral imaging in nondestructive analysis of plant nutrient status.

Nutrient management is vital for plant growth and development. Nitrogen (N) is a crucial nutrient, and its estimation plays a pivotal role in agriculture. Y. Wang, L. Suarez, D. Ryu, and P. J. Zarco-Tejada (2023) found that Solar Induced Chlorophyll Fluorescence (SIF) coupled with chlorophyll content (Cab) has shown promise in estimating leaf N levels. Additionally, the study explores the contribution of other biochemical constituents, such as Cx, in explaining the variability in leaf N levels. The findings highlight the significance of accurate nutrient analysis in agriculture.

An analytical comparison of these models is discussed in the next section of this text.

ANALYTICAL COMPARISON OF THE REVIEWED MODELS

This segment outlines various methods and their objectives, the types of fruits or objects they analyze, the spectral ranges or techniques utilized, and the classification or evaluation metrics used to measure their performance levels. In Table **4**, each method is briefly described to highlight its specific focus and application.

Table 4. Analysis of machine learning models used for hyperspectral processing of image samples.

Authors Name	Objective	Fruits/Objects Analyzed	Spectral Range	Classification Accuracy
A. Kamaru-Zaman	Fig fruit detection using deep learning.	Fig fruits	Deep learning models (YOLOv3, YOLOv4) with 770nm, 840nm, 910nm, and 960nm range LEDs.	YOLOv4 outperformed YOLOv3 in average precision (90.02% *vs.* 81.40%).
A. Vinci	Canopy characterization of hazelnut trees.	Hazelnut trees	UAV technology and the DJI Phantom 4 Multispectral UAV	Good performance in the canopy width and volume estimation
A. Zeb	Fruit classification using NIRS.	11 different fruits	NIR (725-975 nm) and four features at 770nm,840nm, 910nm, 960nm(O-H and C-H overtones)	QDA classifier: 100% (Cross-validation), 93.02% (Test data) for NIR QDA classifier: 97.1% (Cross-validation), 90.38% (Test data) for 4 features.
D. Buonocore	Smart sensing system for the fig drying process.	Figs	Smart sensing system for monitoring fig drying process.	Effective in speeding up the drying process.
E. Kizilay	Freshness assessment of walnuts.	Walnuts	Microwave transmission coefficients (6.8-12.4 GHz).	RMSE for different walnut samples.
J. Wei	Raisin classification using machine learning.	Raisin types	Various features and machine learning models.	Random Forest: Highest recognition success rate.
L. Zabawa *et al.*,	Berry detection in vineyards using CNN.	Grape berries	CNN on images acquired with the Pheno liner.	High accuracy (94.0% for VSP, 85.6% for SMPH).
M. Dejanov	β-carotene content prediction in apricots.	Apricots	VIS-NIR (400–1000 nm) and NIR (900–1700 nm).	R^2_c = 0.91 (Calibration), R^2_v = 0.84 (Validation) RMSE_c = 20.1 (Calibration), RMSE_v = 27.1 (Validation).
M. Greco *et al.*	Hazelnut quality classification with SOM.	Hazelnuts	Frequency range 18 and 40 GHz for Transmission mode utilized.	Distinguishing healthy and rotten hazelnuts.

(Table 4) cont.....

Authors Name	Objective	Fruits/Objects Analyzed	Spectral Range	Classification Accuracy
M. Greco	Hazelnut quality assessment using THz imaging.	Hazelnuts	THz radiation (97 GHz)	Discriminating between healthy and rotten hazelnuts.
N. E. Md Rosli	Fig fruit image segmentation.	Fig fruits	Image segmentation methods (Threshold, K-means clustering, Sharp U-Net).	Sharp U-Net Showed the highest accuracy.
O. Sahin,	Raisin classification using machine learning.	Raisin grains	Morphological features of raisin grains.	Multi-Layer Perceptron: 94.07% (Accuracy).
R. Wang, N. Tuerxun and J. Zheng,	Chlorophyll content estimation in vegetation..	Vegetation	Multi-spectral UAV images at different flight altitudes.	R^2=0.69 to 0.54, RMSE=2.67 to 3.52 (Model Evaluation).
S. Bayrakdar	Hazelnut type and quality determination.	Shelled hazel nuts	Image processing by size and shape characteristics.	84% accuracy for grouping shelled.
Y. Tang	Grape berry Counting using Spatial Attention.	Grape berries	Spatial Attention Grape Berry Counting Network.	Lower counting Error (MAE 35.39, MSE 47.39).

The analysis table compares methods used in different agriculture and food quality assessment studies. These methods employ various technologies to achieve their objectives, from spectroscopy and microwave measurements to machine learning and deep learning models. Table **4** shows that each method targets specific agricultural or food quality-related tasks, such as predicting β-carotene content in apricots, classifying different fruits, estimating chlorophyll content in vegetation, assessing freshness of walnuts, and more. The choice of spectral range or technology is tailored to the unique requirements of each task set.

The classification accuracy or evaluation metrics vary depending on the method's objectives. Some methods achieve high accuracy, such as the QDA classifier for fruit classification using NIRS. In contrast, others focus on specific performance indicators like R^2 and RMSE for the chlorophyll content estimation process. The analysis table comprehensively summarizes various studies on hazelnut processing and quality control. These studies employ diverse methodologies and techniques to address different aspects of hazelnut quality assessment. Here are some key takeaways from the analysis.

Non-Destructive Testing: Methods like TD-NMR, as seen in studies by D. Di Caro, C. Liguori, A. Pietrosanto, and P. Sommella (2018) and (2017), offer non-destructive testing capabilities for hazelnuts. These techniques enable the assessment of hazelnut quality without compromising the integrity of the nuts. **Innovation:** Studies like D. Di Caro (2020) and B. Yalçın and Y. Moazzen (2015) proposed innovative methods, such as spectral clustering and approximate spectral clustering, to address hazelnut-related challenges.

Aflatoxin Contamination: Detecting aflatoxin contamination in hazelnuts is vital for food safety. H. Kalkan and Y. Yardimci (2008) investigated the use of multispectral imaging to identify discriminative bands for aflatoxin contamination detection. **X-Ray Imaging:** I. Khosa and E. Pasero (2014) introduced a method for detecting, segmenting, and classifying hazelnuts using X-ray images. This approach showcases the potential of advanced imaging technologies in quality control.

Remote Sensing: Research by K. Taşdemir and S. Reis (2011) demonstrated the use of remote sensing techniques for detecting hazelnut orchards. This approach leverages high-resolution imagery to enhance orchard detection, thereby aiding in agricultural management.

Data Management: Ş. Aydın, U. Ünal, and M. N. Aydın (2018) emphasized the significance of open data in hazelnut agriculture, highlighting the need for accessible and meaningful data for stakeholders.

Automation and Accuracy: Studies such as M. Demir *et al.* (2012), S. A. Guvenc *et al.* (2015), and I. Onaran focus on automating the detection and classification of hazelnuts. They achieve high classification accuracy, indicating the potential for reducing manual labor and improving accuracy in hazelnut processing.

An analytical review of these methods can be discussed in Table **5** as follows:

Table 5. Machine learning methods used for analysis.

Authors Name	Key Focus	Methodology	Classification Accuracy	Significant Findings
B. Yalçın	Hazelnut Field Detection using Spectral Clustering.	Approximate Spectral Clustering.	N/A	Introduces a novel approach for hazelnut field detection.
D. Di Caro	Inline Quality Evaluation of Inshell Hazelnuts using TD-NMR.	TD-NMR Analysis.	N/A	Focuses on reducing execution time for inline quality evaluation of hazelnuts.

(Table 5) cont.....

Authors Name	Key Focus	Methodology	Classification Accuracy	Significant Findings
D. Di Caro	In-shell Hazelnut Selection Using TD-NMR.	TD-NMR Analysis.	N/A	Develop a method for selecting well-developed hazelnuts using TD-NMR.
D. Di Caro	Quality Assessment of Hazelnuts using TD-NMR.	TD-NMR Technique.	N/A	Utilizes TD-NMR for quality assessment of hazelnuts.
H.Kalkan	Aflatoxin Contamination Detection in Hazelnuts using Multispectral Imaging.	Multispectral Imaging.	N/A	Identifies discriminative bands for aflatoxin contamination in hazelnuts.
I.Khosa	Hazelnut Detection, Segmentation, and Classification Using X-Ray Images.	X-Ray Imaging and Anomaly Detection.	98.6% (Negative Examples).	Utilizes ray images for real-time hazelnut detection and classification.
I. Onaran	Detection of Empty Hazelnuts Using Acoustic Signals.	Feature Analysis and SVM.	98%	Successfully detects empty hazelnuts through acoustic signals.
I. Onaran	Separating Empty and Full Hazelnuts Using Sparse Signal Representation.	Sparse Signal Representation.	98.3%	Achieves high accuracy in separating empty and whole hazelnuts.
K. Taşdemir	Remote Sensing for Hazelnut Orchard Detection.	Rule-based Classification.	N/A	Utilizes remote sensing for the accurate detection of hazelnut orchards.
K. Taşdemir	Detection of Hazelnut Orchards in Remote Sensing Images.	Self-Organizing Map.	N/A	Proposes a method for detecting hazelnut orchards in remote sensing imagery.
M. Demir, M. Ulutaş, and H. Türe,	Automated Detection of Partially Skinne Hazelnuts.	Fuzzy Logic and Image Processing.	N/A	Proposes an automated system for identifying partially skinned hazelnuts.
N. Kumar Naik	Automated Quality Assessment of Almonds.	Image Processing.	N/A	Proposed an automated method for assessing almond quality based on chips and scratches.
Ş. Aydın,	Hazelnut Data Management and Open Data.	Data Processing Model.	N/A	Emphasizes the importance of open data in hazelnut agriculture.
S. A. Guvenc	Classification of Processed Hazelnut Kernels.	Computer Vision Approach.	93.57%	Achieves high classification accuracy for processed hazelnut kernels.

(Table 5) cont.....

Authors Name	Key Focus	Methodology	Classification Accuracy	Significant Findings
Y. H. Habiboğlu	Separation of Shell Pieces From Hazelnut and Pistachio Kernels.	Impact Vibration Analysis.	96.7% (Hazelnuts)	Uses impact vibration analysis for the separation of nut-shell pieces.

These studies collectively contribute to advancing hazelnut processing and quality control, offering a range of techniques that can benefit the hazelnut industry regarding efficiency, accuracy, and safety levels. The following table summarizes and analyzes the methodologies and outcomes of various agriculture and environmental science research papers. Each paper addresses a specific application, and its respective methods and results are assessed based on relevant performance metrics. These findings contribute to advancing agricultural practices, environmental monitoring, and food safety levels. Similar evaluations for deep learning methods are presented in Table **6**.

Table 6. Comparative analysis of deep learning models.

Authors Name	Application	Methodology	Performance Metrics	Key Findings
A. C. Cruz, S. Magana, D. Greco, L. D. Bellis, and A. Luvisi,	Plant Physiology.	Solar Induced Chlorophyll Fluorescence (SIF), Chlorophyll a+b Content (Cab), and Xanthophyll Cycle(Cx).	Contribution of Cx to leaf N-estimation.	Cx ranks third in explaining leaf N variability and enhancing N estimation models.
Divyanth, L.G., Chakraborty, S., Li, B. *et al.*	Disease Detection.	Deep Learning with Convolutional Neural Networks (DenseNet 201, Inception V3, ResNet 101 V2, VGG 19, Xception).	Accuracy (88.72%).	Efficient disease detection in almond trees using deep learning, with potential for cheap and fast diagnosis.
H. R. Seireg	Remote Sensing.	Spatial-Spectral Two-Branch CNN.	Macro Average Recall (0.97).	Improved segmentation of hyperspectral data for class separation.
J. M. Sánchez, *et al*	Agriculture.	Ensemble Machine Learning Algorithms.	R-squared (0.984) and RMSE (179.898).	High-accuracy wild blueberry yield prediction using ensemble machine learning.
K. Susuki *et al.*	Environmental Monitoring.	Energy Balance Models (METRIC, STSEB) Using Landsat 8.	Average Error (< 1.0 mm/day).	Accurate monitoring of evapotranspiration in almond orchards using remote sensing data samples.

(Table 6) cont.....

Authors Name	Application	Methodology	Performance Metrics	Key Findings
K. Susuki *et al.*	Food Processing.	Mechanical Design and Distillation Process Simulation.	Production Ratio (1:7).	Efficient distillation of fermented blueberries for antioxidant-rich beverages.
Nirere *et al.*	Plant Classification	Improved EfficientNet for blueberry variety Classification.	Accuracy (>97%).	High accuracy in classifying blueberry varieties using an improved EfficientNet.
P. B. Cubas *et al.*	Food Safety.	Data Analysis.	Identification of Pesticide Residues.	Identification of causes of pesticide residues in agricultural products and suggestions for improvement.
S. A. Guvenc *et al.*	Crop Management.	YOLOv8 and custom CNN for blueberry detection and ripeness classification.	Accuracy (91.14%)	Successful detection and classification of blueberries based on ripeness levels.
P. B. Cub as Muñoz	Crop Management.	Mask R-CNN and custom CNN for blueberry detection and ripeness classification.	Accuracy (90.74%).	Accurate detection and classification of blueberries based on ripeness levels.
Y. Chao and S. Yan	Agriculture.	Deep Learning Model (Faster-RCNN) for tree identification and counting.	Accuracy (93%)	Successful identification and counting of cashew trees in heterogeneous agricultural areas.
A. Bakiya, V. Neeli, D. M. L. Priya, and C. R. Pavan	Plant Nutrition.	Hyperspectral Imaging and Regression Models (SG-PLS, SVMR, PLS)	RMSE and R-squared.	Accurate non-destructive measurement of leaf nutrition content using a hyperspectral imaging process.

This analysis provides an overview of the methodologies and outcomes discussed in a series of research works related to agriculture and environmental science. Each paper focuses on a specific application, and its methods are evaluated based on relevant performance metrics. The findings from these studies highlight advancements in disease detection, environmental monitoring, plant physiology, crop management, food processing, and other areas for different use cases. These insights contribute to ongoing efforts to improve agricultural practices, enhance food safety, and optimize resource utilization in agriculture and the environment.

CONCLUSION AND FUTURE SCOPE

As presented in the cited papers, the comprehensive analysis of various methods, techniques, and algorithms employed in agricultural and food science sheds light on the diverse approaches to solving critical challenges in these domains. This analytical conclusion delves into the key findings and methodologies presented in the documents, offering insights into their broader implications and potential future directions. The findings and methods discussed in these studies open up several opportunities for future research and development in the field of agricultural and food science:

Integration of Multiple Sensors: Future research can explore data fusion from multiple sensors, such as hyperspectral imaging, acoustic sensors, and remote sensing, to develop more robust and accurate systems for quality assessment, disease detection, and yield prediction.

Automation and Robotics: Automating tasks like fruit detection, counting, and quality assessment through robotics and machine learning can further enhance the efficiency of agricultural processes and reduce labor costs.

Data Analytics and AI: The application of advanced data analytics techniques, including deep learning and ensemble methods, can continue to improve the accuracy and reliability of predictive models for crop yield estimation and quality control.

Sustainable Agriculture: Research into sustainable agricultural practices, including the use of agrarian residues for energy production, can contribute to environmental conservation and resource efficiency levels.

Open Data Initiatives: Emphasizing the importance of open data, as highlighted, can foster collaboration and knowledge sharing in the farmed sector, enabling the development of innovative solutions.

Sensor Miniaturization: Developing compact and affordable sensors for on-site quality assessment and monitoring can empower small-scale farmers and contribute to food security.

In conclusion, the papers discussed in this analysis showcase the diversity and depth of research in agricultural and food sciences. These studies provide valuable insights into quality assessment, environmental monitoring, and yield prediction and pave the way for innovative solutions to address the challenges faced by the agricultural industry. The future holds immense potential for advancements in sensor technology, automation, and data-driven approaches, ultimately driving the transformation of agricultural practices toward greater sustainability and efficiency levels.

REFERENCES

[1] V. Bakiya, Neeli, D. M. L. Priya and C. R. Pavan "Identification of nutrient content of psidium guajava and syzygium cumini leaves using hyperspectral imaging*2nd International Conference on Smart Technologies and Systems for Next Generation Computing (ICSTSN),* pp. 1-6, 2023.Villupuram, India.
[http://dx.doi.org/10.1109/ICSTSN57873.2023.10151662]

[2] C. Cruz, S. Magana, D. Greco, L. D. Bellis and A. Luvisi, "Detection of Almond Leaf Scorch with Artificial Intelligence for the Agriculture Industry", *5th International Conference on Artificial Intelligence for Industries (AI4I)*, Laguna Hills, CA, USA, pp. 1-4, 2022.
[http://dx.doi.org/10.1109/AI4I54798.2022.00007]

[3] A. Kamaruzaman, M. Farid, C. A. A. Izhar, M. Maruzuki, A. S and M. A. Habibi, "Application of Deep Learning for Fig Fruit Detection in The Wild", *2nd International Conference on Emerging Smart Technologies and Applications (eSmarTA)*, Ibb, Yemen, 2022, pp. 1-6.
[http://dx.doi.org/10.1109/eSmarTA56775.2022.9935356]

[4] C. Vinci, Traini, D. Farinelli and R. Brigante, "Assessment of the geometrical characteristics of hazelnut intensive orchard by an Unmanned Aerial Vehicle (UAV, *IEEE Workshop on Metrology for Agriculture and Forestry (MetroAgriFor),* pp. 218-222, 2022.Perugia, Italy.
[http://dx.doi.org/10.1109/MetroAgriFor55389.2022.9964832]

[5] A. Zeb, W.S. Qureshi, A. Ghafoor, and D.O. Sullivan, "Learning fruit class from short wave near infrared spectral features, an ai approach towards determining fruit type", *8th International Conference on Mechatronics and Robotics Engineering (ICMRE),* pp. 193-196, 2022.Munich, Germany.
[http://dx.doi.org/10.1109/ICMRE54455.2022.9734107]

[6] M. Seydou, P. B. Himbane and L. G. Ndiaye, "Determination and comparison of combustion kinetics parameters of peanut shells, cashew nut shells, palm nut shells, and millet stem", *IEEE Multi-conference on Natural and Engineering Sciences for Sahel's Sustainable Development (MNE3SD)*, Ouagadougou, Burkina Faso, pp. 1-7, 2022.
[http://dx.doi.org/10.1109/MNE3SD53781.2022.9723216]

[7] Y.M. Yalçın, and K. Taşdemir, "Extraction of hazelnut fields using Approximate Spectral Clustering Ensemble", *23nd Signal Processing and Communications Applications Conference (SIU),* pp. 640-643, 2015.Malatya, Turkey.
[http://dx.doi.org/10.1109/SIU.2015.7129905]

[8] Buonocore, M. Carratù, G. Ciavolino, D. Di Caro, C. Liguori and A. Pietrosanto,"A smart sensing system for the fig drying process", *IEEE Workshop on Metrology for Agriculture and Forestry (MetroAgriFor)*, Perugia, Italy, pp. 75-79,2022.
[http://dx.doi.org/10.1109/MetroAgriFor55389.2022.9964768]

[9] D. Di Caro, C. Liguori, A. Pietrosanto, and P. Sommella, "Quality Assessment of the Inshell Hazelnuts Based on TD-NMR Analysis", *IEEE Trans. Instrum. Meas.,* vol. 69, no. 6, pp. 3770-3779, 2020.
[http://dx.doi.org/10.1109/TIM.2019.2934662]

[10] D. Di Caro, C. Liguori, A. Pietrosanto and P. Sommella, "Advanced TD-NMR based estimation of the inshell hazelnuts quality parameters", *IEEE International Instrumentation and Measurement Technology Conference (I2MTC),* USA, pp. 1-6.
[http://dx.doi.org/10.1109/I2MTC.2018.8409643]

[11] D. Di Caro, C. Liguori, A. Pietrosanto and P. Sommella, "Using a SVD-based algorithm for T2 spectrum calculation in TD-NMR application to detect hidden defects in hazelnuts", *IEEE International Instrumentation and Measurement Technology Conference (I2MTC)*, Turin, Italy, pp. 1-6, 2017.
[http://dx.doi.org/10.1109/I2MTC.2017.7969849]

[12] L.G. Divyanth, S. Chakraborty, B. Li, D.C. Weindorf, P. Deb, and C.J. Gem, "Non-destructive

prediction of nicotine content in tobacco using hyperspectral image–derived spectra and machine learning", *J. Biosyst. Eng.,* vol. 47, no. 2, pp. 106-117, 2022. [http://dx.doi.org/10.1007/s42853-022-00134-0]

[13] Kizilay, C. Aydinalp and M. N. Akinci, "Analysis of walnut freshness: Preliminary Study, "10th International Conference on Wireless Networks and Mobile Communications (WINCOM), Istanbul, Turkiye, pp. 1-4, 2023. [http://dx.doi.org/10.1109/WINCOM59760.2023.10322950]

[14] Kalkan and Y. Yardimci, "Detection of contaminated hazelnuts by multispectral imaging", *Communication and Applications Conference,* pp. 1-4, 2008.Aydin, Turkey. [http://dx.doi.org/10.1109/SIU.2008.4632687]

[15] H.R. Seireg, Y.M.K. Omar, F.E.A. El-Samie, A.S. El-Fishawy, and A. Elmahalawy, "Ensemble machine learning techniques using computer simulation data for wild blueberry yield prediction", *IEEE Access,* vol. 10, pp. 64671-64687, 2022. [http://dx.doi.org/10.1109/ACCESS.2022.3181970]

[16] Khosa and E. Pasero, "Feature extraction in X-ray images for hazelnuts classification, International Joint Conference on Neural Networks", IJCNN, Beijing, China, pp. 2354-2360, 2014. [http://dx.doi.org/10.1109/IJCNN.2014.6889661]

[17] B. Onaran, Dulek, T. C. Pearson, Y. Yardimci and A. E. Cetin, "Detection of empty hazelnuts from fully developed nuts by impact acoustics, *13th European Signal Processing Conference,* pp. 1-4, 2005.Antalya, Turkey.

[18] N.F. Onaran, Ince, A. H. Tewfik and A. E. Cetin, "A signal representation approach for discrimination between full and empty hazelnuts, *15th European Signal Processing Conference,* pp. 2464-2468, 2007.Poznan, Poland.

[19] M. Sánchez, J. González-Piqueras, J. M. Galve, L. Simón and R. López-Urrea, "Monitoring Water Use in Almond Orchards Through Surface Energy Balance Applied to Landsat Imagery in Southeastern Spain", *IGARSS 2023 - IEEE International Geoscience and Remote Sensing Symposium,* Pasadena, CA, USA, pp. 2811-2814, 2023. [http://dx.doi.org/10.1109/IGARSS52108.2023.10283246]

[20] J. Wei, "Identification of Raisin Varieties Based on Machine Learning Classification Models", *2023 IEEE International Conference on Sensors, Electronics and Computer Engineering (ICSECE),* pp. 1610-1613, 2023.Jinzhou, China. [http://dx.doi.org/10.1109/ICSECE58870.2023.10263345]

[21] A. Susuki, Ipanaque and L. Vinces, "A design of a distillation column to produce an alcoholic beverage based on Blueberries", *Congreso Internacional de Innovación Tendencias en Ingeniería.* [http://dx.doi.org/10.1109/CONIITI61170.2023.10324174]

[22] Taşdemir and S. Reis, "Land cover identification for finding hazelnut fields using WorldView-2 imagery", *IEEE International Geoscience and Remote Sensing Symposium,* pp. 158-161, 2011.Vancouver, BC, Canada. [http://dx.doi.org/10.1109/IGARSS.2011.6048922]

[23] Taşdemir, "Classification of hazelnut orchards by self-organizing maps, IAPR Workshop on Pattern Recognition in Remote Sensing", Istanbul, Turkey, pp. 1-4, 2010. [http://dx.doi.org/10.1109/PRRS.2010.5742803]

[24] L. Zabawa, A. Kicherer, L. Klingbeil, A. Milioto, R. Töpfer, H. Kuhlmann, and R. Roscher, "Detection of single grapevine berries in images using fully convolutional neural networks," in *Proc. 2019 IEEE/CVF Conf. Comput. Vis. Pattern Recognit. Workshops (CVPRW)*, 2019, pp. 2571–2579. [http://dx.doi.org/10.1109/CVPRW.2019.00313]

[25] Dejanov, "Evaluation of β-carotene content in apricots during the drying process using visual and near Infrared spectroscopy", 8th international conference on energy efficiency and agricultural engineering (EE&AE), Ruse, Bulgaria, pp. 1-6, 2022.

[http://dx.doi.org/10.1109/EEAE53789.2022.9831255]

[26] M.U. Demir, and H. Türe, "Fuzzy Logic based identification of partially skinned hazelnuts on a conveyor belt", *International Symposium on Innovations in Intelligent Systems and Applications,* pp. 1-5, 2012.Trabzon, Turkey.
[http://dx.doi.org/10.1109/INISTA.2012.6247043]

[27] Greco, "Selecting hazelnuts by coupling a self-organizing map (SOM) and an experimental system operating in transmission configuration", *48th International Conference on Infrared, Millimeter, and Terahertz Waves (IRMMW-THz),* pp. 1-3, 2023.Montreal, QC, Canada.
[http://dx.doi.org/10.1109/IRMMW-THz57677.2023.10299380]

[28] M. Greco, E. Giovenale, F. Leccese, and A. Doria, "A discrimination of healthy and rotten hazelnuts using a THz imaging scanner", *IEEE Workshop on Metrology for Agriculture and Forestry (MetroAgriFor),* pp. 229-233, 2022.Perugia, Italy.
[http://dx.doi.org/10.1109/MetroAgriFor55389.2022.9964672]

[29] M. Yi, C. Zhao, F. Liao, and W. Yao, "Classification of blueberry varieties based on improved efficientNet", *4th International Academic Exchange Conference on Science and Technology Innovation (IAECST),* pp. 411-415, 2022.Guangzhou, China
[http://dx.doi.org/10.1109/IAECST57965.2022.10062152]

[30] Nirere, A., Sun, J. & Yuhao, Z. ,"A Rapid Non-destructive Detection Method for Wolfberry Moisture Grade Using Hyperspectral Imaging Technology", *J Nondestruct Eval 42*, 2023.
[http://dx.doi.org/10.1007/s10921-023-00944-y]

[31] E. Md Rosli, S. Setumin, A. Nugroho, A. I. Che Ani, M. Ikmal Fitri Maruzuki and M. S. Osman, "Fig fruit image segmentation using threshold, k-means clustering, and sharp u-net techniques", *2nd International Conference on Emerging Smart Technologies and Applications (eSmarTA),* Ibb, Yemen, pp. 1-6, 2022.
[http://dx.doi.org/10.1109/eSmarTA56775.2022.9935411]

[32] K. Naik, P. Kumar Sethy and S. Kumari Behera, "Almond Quality Assessment using Image Processing Techniques,", IEEE 8th International Conference for Convergence in Technology (I2CT), Lonavla, India, pp. 1-4, 2023.
[http://dx.doi.org/10.1109/I2CT57861.2023.10126476]

[33] O. Sahin "Raisin grain classification using machine learning models", *IEEE International Students' Conference on Electrical, Electronics and Computer Science (SCEECS),* Bhopal, India, pp. 1-6, 2023.
[http://dx.doi.org/10.1109/SCEECS57921.2023.10063039]

[34] B. Cubas Muñoz, R.J. Huaman Kemper, and S.R. Prado Gardini, "Artificial vision strategy for Ripeness assessment of Blueberries on Images taken during Pre-harvest stage in Agro industrial Environments using Deep Learning Techniques", *IEEE XXX International Conference on Electronics, Electrical Engineering and Computing (INTERCON),* pp. 1-6, 2023.Lima, Peru
[http://dx.doi.org/10.1109/INTERCON59652.2023.10326058]

[35] E. Brayam Cubas Muñoz, Martin Fiestas Sorogastua and S. Ricardo Prado Gardini, "Detection and Classification of ventura-blueberries in five levels of ripeness from images taken during pre-harvest stage using Deep Learning techniques". IEEE ANDESCON: Barranquilla, Colombia, 2022, pp. 1-6.
[http://dx.doi.org/10.1109/ANDESCON56260.2022.9989578]

[36] Wang, N. Tuerxun and J. Zheng, "Scaling effects of chlorophyll content in walnut leaves estimations with coupling Boruta algorithm and machine learning model", *11th International Conference on Agro-Geoinformatics (Agro-Geoinformatics),* Wuhan, China, 2023, pp. 1-5.

[37] Ş. Aydın, U. Ünal, and M.N. Aydın, "Open Data in Agriculture: Sustainable Model Development for Hazelnut farms using semantics", *6th International Conference on Control Engineering & Information Technology (CEIT),* pp. 1-6, 2018.Istanbul, Turkey.
[http://dx.doi.org/10.1109/CEIT.2018.8751875]

[38] S.A. Guvenc, F.A. Senel, and B. Cetisli, "Classification of processed hazelnuts with computer vision

Signal", *Processing and Communications Applications Conference (SIU),* pp. 1362-1365, 2015.Malatya, Turkey.
[http://dx.doi.org/10.1109/SIU.2015.7130094]

[39] B. Bayrakdar, "Çomak, D. Başol and İ. Yücedag, "Determination of type and quality of hazelnut using image processing techniques", *23nd Signal Processing and Communications Applications Conference (SIU),* pp. 616-619, 2015.Malatya, Turkey.
[http://dx.doi.org/10.1109/SIU.2015.7129899]

[40] R. Bayala, S. Malo and Z. Kinda, "DL4Cashew: Deep Learning Model For Cashew Trees Counting", *IEEE International Conference on Contemporary Computing and Communications (InC4)*, Bangalore, India, pp. 1-7, 2023.
[http://dx.doi.org/10.1109/InC457730.2023.10262859]

[41] Y. Pestel-Schiller, Yang and J. Ostermann, " Semantic Segmentation Of Natural And Man- Made Fruits Using A Spatial-Spectral Two-Branches-Cnn For Sparse Data, *12th Workshop on Hyperspectral Imaging and Signal Processing: Evolution in Remote Sensing (WHISPERS),* pp. 1-5, 2022.Rome, Italy.
[http://dx.doi.org/10.1109/WHISPERS56178.2022.9955066]

[42] Y. Chao and S. Yan, "Study on Rapid Detection Model of Pesticide Residues on Fruit Surface Based on Hyperspectral Imaging Technology", *2nd International Conference on Networking, Communications and Information Technology (NetCIT),* pp. 294-297, 2022.Manchester, United Kingdom.
[http://dx.doi.org/10.1109/NetCIT57419.2022.00076]

[43] Y.H. Habiboğlu, R.A. Sevimli, A.E. Çetin, and T.C. Pearson, "Separating nut-shell pieces from hazelnuts and pistachio kernels using impact vibration analysis", *21st Signal Processing and Communications Applications Conference (SIU),* pp. 1-4, 2013.Haspolat, Turkey.
[http://dx.doi.org/10.1109/SIU.2013.6531171]

[44] Y. Tang, Y. Li, M. Sun, and Y. Qi, "Field grape berry counting algorithm based on spatial attention mechanism", *China Automation Congress (CAC),* pp. 2211-2216, 2022.Xiamen, China.
[http://dx.doi.org/10.1109/CAC57257.2022.10055368]

[45] Y. Wang, L. Suarez, D. Ryu, and P.J. Zarco-Tejada, "Evaluating the contribution of Cx to leaf nitrogen quantification using fluspect and airborne imaging spectroscopy in almond orchards", *IGARSS IEEE International Geoscience and Remote Sensing Symposium,* pp. 2783-2786, 2023.Pasadena, CA, USA.
[http://dx.doi.org/10.1109/IGARSS52108.2023.10281716]

CHAPTER 3

Deep Learning Fusion: Potato Leaf Disease Classification through Generative Diffusion Model

Virendra Kumar Shrivastava[1,*], **Prakruthi Ganiga**[1], **Rathnakar Achary**[1], **Chetan J. Shelke**[1], **A. Ezil Sam Leni**[1] and **Ramalakshmi K.**[1]

[1] *Centre of Excellence in Computer Vision, Department of Computer Science and Engineering, School of Advanced Computing, Alliance University, Bangalore, Karnataka-562106, India*

Abstract: Potatoes are one of the most highly consumed foods in the world. Due to several environmental and climate changes, potato leaf diseases are emerging frequently, causing severe damage to potato plants and drastically reducing potato crop yields, which results in financial losses to farmers. This research presents a deep learning generative approach to enhance the accuracy of identifying classical diseases in potato leaves using a novel data augmentation method, specifically the diffusion model. By employing diffusion models, researchers created images resembling real ones showing diseases such as Early Blight, Late Blight, and healthy leaves on potato leaves. These generated images are then combined with others to strengthen the training of a deep learning model, enabling it to better distinguish between early blight, late blight, and healthy leaves. This combination of synthetic data enhances the model's robustness. Demonstrates promising advancements in analyzing potato leaf disease. The significance of this study lies in addressing challenges related to data availability while improving the model's ability to generalize across the stages of the disease. The outcomes have implications for agriculture by providing effective disease-identification methods for potato farmers. The proposed CNN model delivered classification accuracy (98.99%) with ResNet18 integrating with the diffusion model on the potato leaf disease image dataset. Integrating diffusion models represents a step towards optimizing crop management and maximizing yields in potato cultivation.

Keywords: Artificial neural network, Credibility limits, Loan approval, Loan prediction, Machine learning.

INTRODUCTION

The potato, scientifically known as *Solanum tuberosum*, is a crop that serves as a food source for billions of people worldwide. In the industry, it is challenging to safeguard potato plants from diseases that can significantly impact their yield and

* **Corresponding Author Virendra Kumar Shrivastava:** Centre of Excellence in Computer Vision, Department of Computer Science and Engineering, School of Advanced Computing, Alliance University, Bangalore, Karnataka-562106, India; E-mail: Virendra.shrivastava@alliance.edu.in

the quality of the potatoes grown. Identifying and classifying diseases in potato leaves at an early stage is crucial for implementing effective disease management strategies. Recent technological advancements in learning techniques have resulted in automating and improving the accuracy of disease classification in plants. Plant diseases, including those that affect potato leaves, pose a threat to food security. Early blight disease (caused by *Alternaria solani*) and late blight disease in potatoes (caused by Phytophthora infections) are known culprits that negatively affect the health of potato crops [1, 2]. In agriculture, the process of identifying diseases traditionally involves expert inspection. However, this approach is both time-consuming and prone to error. It has become crucial to incorporate technologies to overcome these challenges.

Machine learning is a popular technique for identifying and classifying diseases [3 - 6]. Deep learning, which falls under the umbrella of intelligence, has succeeded in various tasks related to image classification [7 - 10]. One type of learning architecture, called Convolutional Neural Networks (CNNs), has demonstrated the ability to recognize patterns and characteristics within images [11 - 14]. This makes CNNs highly suitable for classifying plant diseases based on leaf images [15, 16].

However, the efficacy of the learning model is primarily dependent on the availability of diverse datasets for training. A significant obstacle in potato leaf disease classification is the scarcity of labelled datasets encompassing different illnesses. Creating and analyzing a well-labeled collection of potato leaf diseases is a challenging and time-consuming task. The limitation [17] dramatically impedes training robust and widely applicable deep learning models. Scientists have explored various data augmentation strategies to address this issue, which involve artificially modifying the training samples to generate a larger dataset. Standard techniques in picture manipulation include flipping, scaling, and rotation. Although these tactics can enhance the generalization capabilities of models to some degree, they may not be sufficient to accurately reflect the complex variations that exist in real-world scenarios [18]. Generative models for data augmentation have made substantial progress thanks to deep learning. Artificial models have been used to create novel products that mimic the functions of living organisms. The goal is to enhance the ease of training deep learning models.

A Generative Adversarial Network is a generative model commonly used to enhance data. It consists of two networks: the generator and discriminator networks. Both components undergo simultaneous training [19, 20]. While the generator creates synthetic data, the discriminator distinguishes between actual samples and fake ones. Synthetic data produced by this adversarial training method is of exceptionally high quality and closely reflects real-world examples [21, 22]. While GANs have shown promise in style transfer and image generation, they may struggle to maintain the authenticity and diversity of the produced data. Synthetic visuals must be accurate and applicable for the model training to be efficient in plant disease classification [23].

In this specific case, generative diffusion models are particularly effective. Drawing inspiration from the physical processes of diffusion, generative diffusion models offer a novel approach to enhancing data. To simulate a diffusion-like effect, the input pictures are gradually exposed to controlled noise during the diffusion process. Deliberately adding noise to synthetic samples makes them appear more realistic, creating several copies of the original images [24 - 26]. Building upon the current foundation of generative diffusion models, this study aims to address the unique challenges of disease classification in potato leaves. The deep learning model must be able to distinguish between Late blight and Early blight, for example, solely by examining the symptoms. This study aims to use generative diffusion models to improve the model's resilience and accuracy in illness classification. Given the dearth of data regarding the classification of diseases affecting potato leaves, our approach carries significant implications for this domain. Generative diffusion models are implemented to augment crop management, increase crop yield, and ensure the long-term viability of the potato farming ecosystem.

RELATED WORK

This study introduces Gaussian noise of varying intensities into training images, enhancing data classification. This approach improves the model's generalizability in classifying disease in the images to encompass robust feature learning. Including chaotic inputs during the training enhances the accuracy and robustness of the model. The diffusion model is indispensable for enhancing accuracy through effective data augmentation and robust feature capture.

Researchers have previously employed datasets, such as PlantVillage and Mendely, to train models and investigate the effects of different potential diseases on potato plants, particularly potato leaf disease. Data augmentation techniques, *viz* a *viz* horizontal rotation, shear, and zoom, were applied on potato leaves to meet the variability in the dataset. K-means segmentation achieved an accuracy of 97% for K=3 [27]. Shin D., Kim T., and Min B. employed a data augmentation model for image-to-image translation. The PlantVillage dataset was used to train the Cycle GAN, which attained an accuracy of 97.74%, after using various deep learning approaches [28]. Among these, the ResNet18 convolutional network demonstrates the highest level of effectiveness. Barman U., Sahu D., and Das J.

devised the Self-Build CNN (SBCNN) method for detecting diseases in potato leaves. The SBCNN evaluates images both with and without augmentation. The highest validation accuracy achieved for augmented images is 96.75%, while for non-augmented images, it is 96.98% [29]. Pandian J. A and Annette B utilized an extensive range of image augmentation methods, which encompassed flipping, cropping, and rotation, along with some color transformation, PCA color augmentation, noise injection, and advanced techniques like Generative Adversarial Networks (GANs) and Neural Style Transfer (NST). According to their claim, combining deep learning, colour augmentation, and positional augmentation datasets results in better classification performance compared to other datasets, with an accuracy of approximately 92% [30]. V. K. Shrivastava (2023) developed a model that utilizes a CNN architecture, incorporating data augmentation, transfer learning, and hyperparameter optimization. Using the PlantVillage and PDLD datasets from Kaggle, the model accomplishes a remarkable 99.2% accuracy after 50 epochs [31].

Data augmentation approaches were integrated into the deep learning model meant to classify potato leaf illnesses (Table **1**).

Table 1. Summary of deep learning model.

Algorithm	Dataset	Accuracy	References
ODCNN	PDLD	99.2	[32]
ResNet50	Plant Village	98%	[33]
VGG19	Plant Village	91%	[33]
CNN	PLD	98.07%	[34]
PDDCNN	Plant Village	96.71	[35]

METHODOLOGY

Dataset

The Potato Disease Leaf Dataset, comprising 4072 images of potato leaves categorized into three groups, served as the basis for this investigation. Kaggle was the source of the dataset. Inside the main folder called "PLD_3_Classes_256," the dataset was divided into three subfolders: "Testing," "Training," and "Validation." Early Blight, Healthy, and Late Blight were the folders of the "Testing" subfolder. There were 162, 102, and 141 images in each group. The "Training" subfolder follows a similar pattern, with the same three groups and corresponding numbers of photographs (1303, 816, and 1132). There were three subfolders in the "Validation" folder: "Early Blight," "Healthy," and

"Late Blight." The "Early Blight" folder contained 163, the "Healthy" folder contained 102, and the "Late Blight" folder contained 151 photographs. The balanced representation of classes in the testing, validation, and training sets was achieved by this dataset split, which expedited model training and testing. "Early blight," "Late blight," and "healthy" were naming standards that accurately reflected the state of the potato leaves. The information is well-structured and follows industry standards, enabling thorough research and evaluation of diseases affecting potato leaves. The number of pictures used in the potato leaves image dataset is concisely described in Table **2**.

Table 2. Potato leaf disease dataset sample size for train-validation-test split.

Category	Train	Test	Valid	Total No. of Images
Early Blight	1303	162	163	1628
Late Blight	1132	141	151	1424
Healthy	816	102	102	1020

Data Preprocessing

Preparing the input data to improve picture quality for effective neural network training was an integral part of this study. During the transformation step of the process, the dataset's original raw images were resized to a standard format of 224x224 pixels. Due to this, the preprocessing photos maintained a uniform size. The images were then transformed into tensors, an appropriate data structure for deep learning calculations. As a result, the neural network could better comprehend and analyze the picture data. The next step was normalization, which transformed the pixels' intensity values to fit a predetermined scale, often with a mean value set to 0 and a standard deviation value set to 1. Standardization improved the consistency and effectiveness of the training process.

Generating datasets for testing, validation, and training purposes is recommended using the pre-processed images. Labels designated the class or group to which a particular image was assigned, and each dataset comprised images accompanied by such descriptions. A data loader was employed to import and manage these datasets efficiently. Furthermore, data augmentation was essential to this study, particularly during the training of the diffusion model. Simulating various training conditions with controlled random noise in the images enhanced the neural network's adaptability to multiple changes in the input data. Additionally, the method detected the presence of a GPU; when present, the model leveraged parallel processing to enhance the training process through GPU acceleration.

Data Augmentation with Diffusion Model

This study highlights the need for data augmentation on potato images *via* the diffusion model to improve the accuracy and dependability of the image classification model. A prevalent method for enhancing training involves incorporating controlled noise that simulates real-life situations [36 - 38]. The diffusion model enhances its precision by manipulating the quantity of additional noise added to the input images, denoted as betas and drift coefficients, which are generated to carry out data augmentation [39].

This enhancement to the model is crucial for increasing its adaptability across various contexts and reducing the likelihood of overfitting. Improving the diversity of the potato leaf images in the training data contributes to the neural network model's improved accuracy in detecting patterns amidst interference [40]. This research demonstrates the pragmatic benefits of this technology, specifically enhanced precision, mainly when the model deals with noisy input images. The diffusion model incorporates data augmentation, primarily responsible for the model's strength and effectiveness in image classification (refer to Fig. **1**).

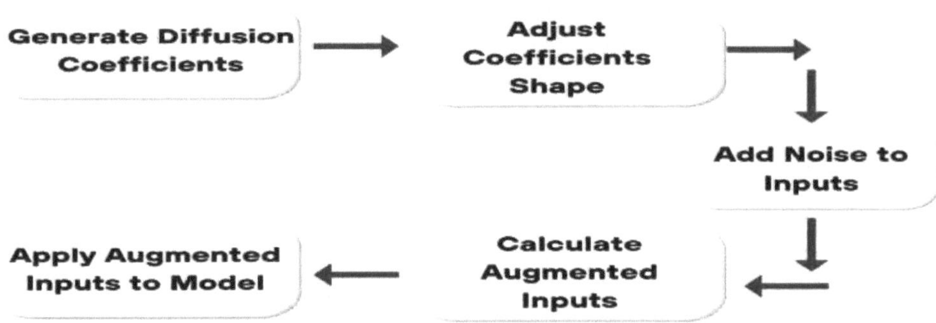

Fig. (1). Diffusion model for data augmentation.

Deep Learning Model for Classification of Images

The performance of the ResNet-18 image classification model is notably exceptional due to its reliance on the ResNet-18 architecture, which is widely recognized as a convolutional neural network (CNN). Various classifications of potato plant leaf images, including those affected by late blight, early blight, and healthy leaves [41], are implemented by employing the ResNet-18 model. The strength of the model resides in its profound architecture. It is fortified by residual connections and its ability to interpret intricate hierarchical patterns within many images. The updated ResNet-18 model incorporates a fully connected layer to generate greater specificity. It is optimized to correspond with the number of classes in the image training dataset and ensures seamless integration into the image classification procedure.

The Adam optimizer [42], which generally has a learning rate of 0.0001 and a weight decay of 1e-4, is utilized to configure the model. The cross-entropy loss function is employed during the training phase. It is highly effective for image classification tasks involving multiple classes. To enhance the learning process, a StepLR scheduler is implemented at every five epochs to decrease the learning rate of the model by a factor of 0.5. Parameter selection is employed to mitigate the risk of overfitting and enable the model to adapt to the intricacies of the dataset. The proposed model (refer to Fig. **2**) effectively classifies images of potato leaf disease identification through a modified ResNet-18 architecture.

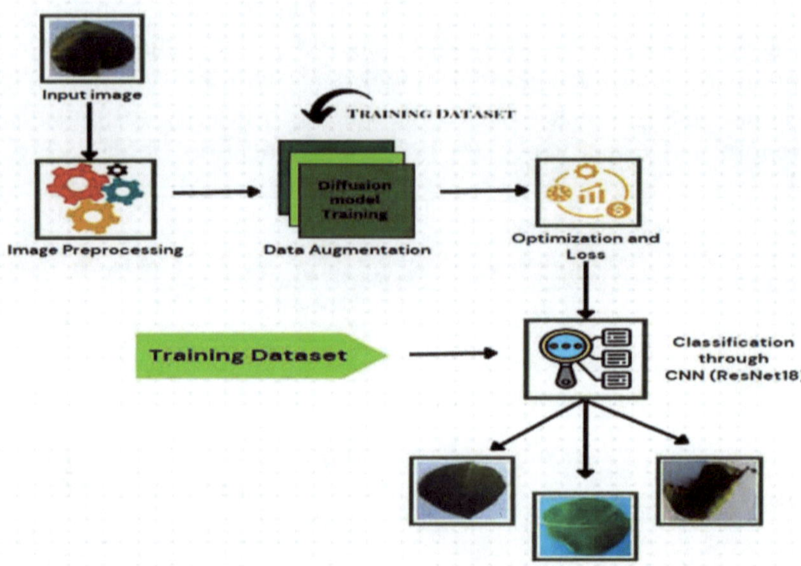

Fig. (2). Proposed model.

RESULT AND DISCUSSION

The primary objective of this research is to classify diseased potato leaf images into three distinct categories: early blight, late blight, and healthy leaves. The study employed deep learning techniques, specifically a modified ResNet 18 (CNN) model. This work utilized a potato image dataset of 4072 images, comprising 1628 early blight, 1424 late blight, and 1020 healthy potato leaf images. A further dataset split is performed on the images to facilitate training, testing, and model validation. Fig. (**3**) depicts Training Loss *versus* Validation Accuracy, and Fig. (**4**) presents Training Loss *versus* Validation Loss.

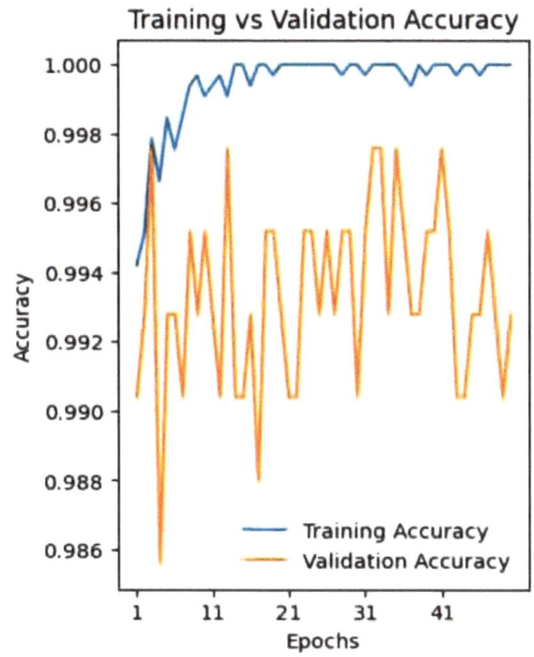

Fig. (3). Training *versus* validation accuracy.

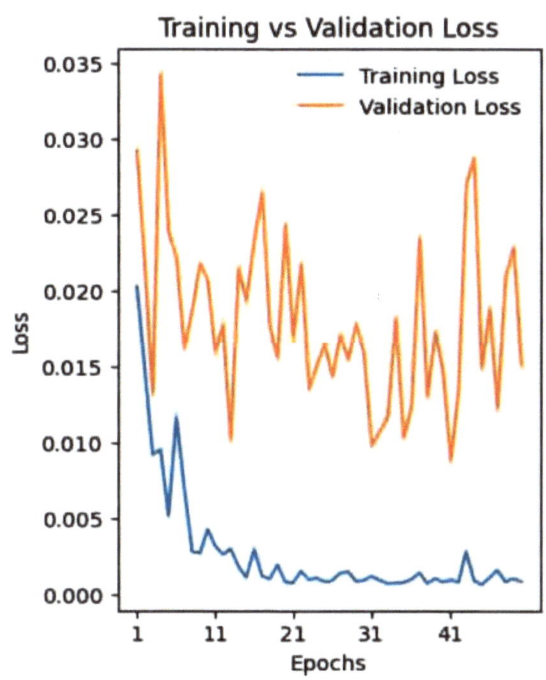

Fig. (4). Training *versus* validation loss.

The ResNet 18 model was trained using this dataset, which was enriched during training using the diffusion model for augmenting the potato leaves image dataset. The resilience and adaptability of the diffusion model were enhanced through the integration of controlled noise. The Adam optimizer, CrossEntropyLoss, and a StepLR scheduler were implemented to optimize learning during training. The confusion matrix in the matplotlib library represents the performance indicators of the model's testing set evaluation.

By integrating the diffusion model into the ResNet 18 architecture, this methodology demonstrates potential improvement in the classification of diseases in potato leaves with an accuracy of 98.99%. This chapter makes a significant contribution to plant disease classification by incorporating data augmentation methods to enhance the flexibility and precision of models.

CONCLUSION

This chapter efficiently classified disease on potato leaves by employing a convolutional neural network model. The model classified late blight, early blight, and healthy leaves. Integrating a customized ResNet 18 architecture with a diffusion model for data augmentation for classification resulted in enhanced feature extraction. A framework to strengthen disease categorization has been developed. A noticeable performance improvement is observed when using the proposed model, which utilizes a diffusion model to enhance controlled data. The CNN model delivered classification accuracy (98.99%) with ResNet 18 integrating with the diffusion model onto the potato leaf disease image dataset. By employing diffusion model data augmentation techniques, the model produced resilient predictions regardless of fluctuations or noise in the input data for potato leaf images. To advance this study, future endeavors should aim to enhance the accessibility of disease detection in potato leaves by developing a user-friendly application. This application would include the optimized Convolutional Neural Network (CNN) model mentioned earlier, enabling farmers to capture leaf photographs using their smartphones. Afterward, the model will detect disease in potato leaves. Provide recommendations for implementing disease management measures.

This proactive initiative aims to provide farmers with resources that enable them to identify diseases, increase crop yield, and promote sustainable agricultural practices.

REFERENCES

[1] L.J. Stridh, H. Mostafanezhad, C.B. Andersen, F. Odilbekov, L. Grenville-Briggs, Å. Lankinen, and E. Liljeroth, "Reduced efficacy of biocontrol agents and plant resistance inducers against potato early blight from greenhouse to field", *J. Plant Dis. Prot.,* vol. 129, no. 4, pp. 923-938, 2022.

[http://dx.doi.org/10.1007/s41348-022-00633-4]

[2] T. Nazir, M.M. Iqbal, S. Jabbar, A. Hussain, and M. Albathan, "EfficientPNet—An optimized and efficient deep learning approach for classifying disease of potato plant leaves", *Agriculture,* vol. 13, no. 4, p. 841, 2023.
[http://dx.doi.org/10.3390/agriculture13040841]

[3] R. Batra, M. Mahajan, V. K. Shrivastava, and A. K. Goel, "Detection of COVID-19 using textual clinical data: a machine learning approach," 2021.
[http://dx.doi.org/10.1007/978-981-16-2786-6_5]

[4] M.M. Rao, V.K. Shrivastava. Predicting Probable "Product swaps in customer behaviour: An in-depth analysis of forecasting techniques, factors influencing decisions, and implications for business strategies", *EAI Endorsed Transactions on Scalable Information Systems.* 2023 Oct 3;10(6).

[5] A. Kumar, V. K. Shrivastava, P. Kumar, A. Kumar, and V. Gulati, "Predictive and experimental analysis of forces in die-less forming using artificial intelligence techniques". *Proceedings of the Institution of Mechanical Engineers, Part E: Journal of Process Mechanical Engineering*, 2024.
[http://dx.doi.org/10.1177/09544089241235473]

[6] Swain, A., Swain, P. K., & Satpathy, S. "Smart Farming for Sustainable Development–A Study at Sonepur District of Odisha". *In AI in Agriculture for Sustainable and Economic Management* (pp. 174-178). CRC Press.
[http://dx.doi.org/10.1201/9781003451648-13]

[7] R. Achary, C. J. Shelke, V. K. Shrivastava, P. M. Paul, S. Konda, and M. Billa, "Depression detection based on NLP and ML techniques using text and speech recognition," 2024.
[http://dx.doi.org/10.1007/978-981-99-9436-6_25]

[8] V. K. Shrivastava, A. Kumar, A. Shrivastava, A. Tiwari, K. Thiru, and R. Batra, "Study and Trend Prediction of Covid-19 cases in India using Deep Learning Techniques," in *Journal of Physics: Conference Series*, 2021.
[http://dx.doi.org/10.1088/1742-6596/1950/1/012084]

[9] R. Batra, V. K. Shrivastava, and A. K. Goel, "Anomaly Detection over SDN Using Machine Learning and Deep Learning for Securing Smart City," in *Green Internet of Things for Smart Cities*, 2021.
[http://dx.doi.org/10.1201/9781003032397-13]

[10] K. Lalli, V.K. Shrivastava, and R. Shekhar, "Detecting copy move image forgery using a deep learning model: A review", *2023 International Conference on Artificial Intelligence and Applications, ICAIA 2023 and Alliance Technology Conference, ATCON-1 2023 - Proceeding,* 2023.
[http://dx.doi.org/10.1109/ICAIA57370.2023.10169568]

[11] V.K. Shrivastava, A. Shrivastava, N. Sharma, S.N. Mohanty, and C.R. Pattanaik, "Deep learning model for temperature prediction: an empirical study", *Model. Earth Syst. Environ.,* vol. 9, no. 2, pp. 2067-2080, 2023.
[http://dx.doi.org/10.1007/s40808-022-01609-x]

[12] A. Singhal, M. Phogat, D. Kumar, A. Kumar, M. Dahiya, and V.K. Shrivastava, "Study of deep learning techniques for medical image analysis: A review", *Mater. Today Proc.,* vol. 56, pp. 209-214, 2022.
[http://dx.doi.org/10.1016/j.matpr.2022.01.071]

[13] Tiwari, D., Ashish, M., Gangwar, N., Sharma, A., Patel, S., & Bhardwaj, S. (2020, May). Potato leaf diseases detection using deep learning. In *2020 4th international conference on intelligent computing and control systems (ICICCS)* (pp. 461-466). IEEE.
[http://dx.doi.org/10.1109/ICICCS48265.2020.9121067]

[14] A. J, J. Eunice, D.E. Popescu, M.K. Chowdary, and J. Hemanth, "Deep learning-based leaf disease detection in crops using images for agricultural applications", *Agronomy (Basel),* vol. 12, no. 10, p. 2395, 2022.
[http://dx.doi.org/10.3390/agronomy12102395]

[15] V. Saini, N. Rai, N. Sharma, and V.K. Shrivastava, "A convolutional neural network based prediction model for classification of skin cancer images", In: *Lecture Notes of the Institute for Computer Sciences, Social-Informatics and Telecommunications Engineering.* LNICST, 2023. [http://dx.doi.org/10.1007/978-3-031-35078-8_9]

[16] R. Mahum, H. Munir, Z-U-N. Mughal, M. Awais, F. Sher Khan, M. Saqlain, S. Mahamad, and I. Tlili, "A novel framework for potato leaf disease detection using an efficient deep learning model", *Hum. Ecol. Risk Assess.,* vol. 29, no. 2, pp. 303-326, 2023. [http://dx.doi.org/10.1080/10807039.2022.2064814]

[17] C. Shorten, and T.M. Khoshgoftaar, "A survey on Image Data Augmentation for Deep Learning", *J. Big Data,* vol. 6, no. 1, p. 60, 2019. [http://dx.doi.org/10.1186/s40537-019-0197-0]

[18] S. Yang, W. Xiao, M. Zhang, S. Guo, J. Zhao, and F. Shen, "Image data augmentation for deep learning: A survey," arXiv preprint arXiv:2204.08610, Apr. 19, 2022.

[19] S. Motamed, P. Rogalla, and F. Khalvati, "Data augmentation using Generative Adversarial Networks (GANs) for GAN-based detection of Pneumonia and COVID-19 in chest X-ray images", *Informatics in Medicine Unlocked,* vol. 27, p. 100779, 2021. [http://dx.doi.org/10.1016/j.imu.2021.100779] [PMID: 34841040]

[20] A. Biswas, Md A. Al, A. Imran, A. T. Sejuty, F. Fairooz, S. Puppala, and S. Talukder, "Generative adversarial networks for data augmentation," in *Data Driven Approaches on Medical Imaging*, Cham: Springer Nature Switzerland, Oct. 17, 2023, pp. 159-177. [http://dx.doi.org/10.1007/978-3-031-47772-0_8]

[21] A. Figueira and B. Vaz, "Survey on Synthetic Data Generation, Evaluation Methods and GANs," 2022. [http://dx.doi.org/10.3390/math10152733]

[22] S. Li, J. Wu, X. Xiao, F. Chao, X. Mao, and R. Ji, "Advances in Neural Information Processing Systems", In: *Revisiting Discriminator in GAN Compression: A Generator-discriminator Cooperative Compression Scheme.*, 2021.

[23] Y. Liao, and Y. Huang, "Deep Learning-Based Application of Image Style Transfer", *Math. Probl. Eng.,* vol. 2022, pp. 1-10, 2022. [http://dx.doi.org/10.1155/2022/1693892]

[24] M. F. Burg, F. Wenzel, D. Zietlow, M. Horn, O. Makansi, F. Locatello, and C. Russell, "Image retrieval outperforms diffusion models on data augmentation," arXiv preprint arXiv:2304.10253, Apr. 2023.

[25] Z. Chen, "Diffusion Models-based Data Augmentation for the Cell Cycle Phase Classification," in *Journal of Physics: Conference Series*, 2023. [http://dx.doi.org/10.1088/1742-6596/2580/1/012001]

[26] B. Trabucco, K. Doherty, M. Gurinas, and R. Salakhutdinov, "Effective data augmentation with diffusion models," arXiv preprint arXiv:2302.07944, Feb. 2023.

[27] M. A. R. Nishad, M. A. Mitu, and N. Jahan, "Predicting and classifying potato leaf disease using K-means segmentation techniques and deep learning networks," in *Procedia Computer Science*, 2022. [http://dx.doi.org/10.1016/j.procs.2022.11.006]

[28] B. Min, T. Kim, D. Shin, and D. Shin, "Data Augmentation Method for Plant Leaf Disease Recognition", *Appl. Sci. (Basel),* vol. 13, no. 3, p. 1465, 2023. [http://dx.doi.org/10.3390/app13031465]

[29] U. Barman, D. Sahu, G.G. Barman, and J. Das, "Comparative assessment of deep learning to detect the leaf diseases of potato based on data augmentation", *2020 International Conference on Computational Performance Evaluation,* 2020. [http://dx.doi.org/10.1109/ComPE49325.2020.9200015]

[30] J. Arun Pandian, G. Geetharamani, and B. Annette, "Data Augmentation on Plant Leaf Disease Image Dataset Using Image Manipulation and Deep Learning Techniques", *Proceedings of the 2019 IEEE 9th International Conference on Advanced Computing, IACC 2019,* 2019. [http://dx.doi.org/10.1109/IACC48062.2019.8971580]

[31] L. Perez, "The effectiveness of data augmentation in image classification using deep learning," arXiv preprint arXiv:1712.04621, 2017.

[32] V.K. Shrivastava, C.J. Shelke, A. Shrivastava, S.N. Mohanty, and N. Sharma, "Optimized Deep Learning Model for Disease Prediction in Potato Leaves", *EAI Endorsed Trans. Pervasive Health Technol.,* vol. 9, no. 1, 2023. [http://dx.doi.org/10.4108/eetpht.9.4001]

[33] Y. Wang, Y. Zhao, J. Zhang, Y. Qian, J. Huang, Q. Zhao, Z. Zhao, and J. Zhao. "DiseSniper: A potato disease identification system based on the ResNet model," in 2022 *10th International Conference on Agro-Geoinformatics, Agro-Geoinformatics* 2022, 2022. [http://dx.doi.org/10.1109/Agro-Geoinformatics55649.2022.9859214]

[34] P. Khobragade, A. Shriwas, S. Shinde, A. Mane, and A. Padole, "Potato leaf disease detection using CNN," in 2022 *International Conference on Smart Generation Computing, Communication and Networking (SMART GENCON)*, Dec. 2022, pp. 1–5. [http://dx.doi.org/10.1109/SMARTGENCON56628.2022.10083986]

[35] J. Rashid, I. Khan, G. Ali, S.H. Almotiri, M.A. AlGhamdi, and K. Masood, "Multi-level deep learning model for potato leaf disease recognition", *Electronics (Basel),* vol. 10, no. 17, p. 2064, 2021. [http://dx.doi.org/10.3390/electronics10172064]

[36] M.A. Diputra, *Potato leaf disease dataset.* Kaggle, 2021.

[37] A. Ayton and D. Nelson, "Diffusion models as data augmentation for medical imaging," 2023.

[38] X. Yu, G. Li, W. Lou, S. Liu, X. Wan, Y. Chen, and H. Li. "Diffusion-Based Data Augmentation for Nuclei Image Segmentation," in Lecture Notes in Computer Science (including subseries Lecture Notes in Artificial Intelligence and Lecture Notes in Bioinformatics), 2023. [http://dx.doi.org/10.1007/978-3-031-43993-3_57]

[39] K. Alomar, H.I. Aysel, and X. Cai, "Data Augmentation in Classification and Segmentation: A Survey and New Strategies", *J. Imaging,* vol. 9, no. 2, p. 46, 2023. [http://dx.doi.org/10.3390/jimaging9020046] [PMID: 36826965]

[40] L. Ma, Y. Hu, Y. Meng, Z. Li, and G. Chen, "Multi-Plant Disease Identification Based on Lightweight ResNet18 Model", *Agronomy (Basel),* vol. 13, no. 11, p. 2702, 2023. [http://dx.doi.org/10.3390/agronomy13112702]

[41] M.A. Hanif, and H. Kaur, "Plant disease detection with Finetuned-ResNet18 for several plants like Tomato, Grape, Orange, Soybean, Squash, Potato, Corn (maize), Strawberry", *Res Mil. (Fr.),* vol. 13, no. 3, pp. 1703-1715, 2023.

[42] Z. Zhang, "Improved Adam Optimizer for Deep Neural Networks," in *2018 IEEE/ACM 26th International Symposium on Quality of Service*, IWQoS 2018, 2019. [http://dx.doi.org/10.1109/IWQoS.2018.8624183]

CHAPTER 4

Fuzzification for Precision Farming with Minimal Human Intervention

Sudheer Mangalampalli[1,*], **Ganesh Reddy Karri**[2], **Pelluru Pavan Kumar Reddy**[2], **Ramith Yadav S.**[2], **K. Varada Rajkumar**[3] and **Kiran Sree Pokkuluri**[4]

[1] *Department of CSE, Manipal Institute of Technology, Manipal Academy of Higher Education, Manipal, India*

[2] *SCOPE, VIT-AP University, Andhra Pradesh, India*

[3] *Department of CSE (AI&ML), MLR Institute of Technology, Telangana, India*

[4] *Department of CSE, Shri Vishnu Engineering College for Women, Bhimavaram, India*

Abstract: Seamless communication between humans and things like cars and other such machines in important industry sectors is constantly evolving as each day passes. In the 21st century, crucial advancements in several fields have implemented IoT in them because of its numerous merits. When IoT is put into practice alongside Cloud Computing and Data Analytics, it can reduce the gap and enhance cooperation between the physical and digital worlds, which is essential for sustainability in this hyperconnected era. The project involves collecting data through sensors in the field, which is then sent to the IoT analytics and cloud-based platform ThingSpeak. This platform enables data ingestion and storage, allowing farmers to remotely visualize and analyze real-time data in the form of data streams and take action accordingly. Fuzzy logic is one such approach to computing that works on degrees of truth rather than single-valued Boolean logic, thus yielding better and more accurate conclusions for an imprecise spectrum of data. Fuzzy logic is implemented in this project to control water pumping time based on user-defined thresholds. With an increasing population and depleting resources, precision farming has proven to be efficient as it focuses on minimizing waste and is cost- and energy-efficient. The objective of this paper is to develop a system that assists farmers in remotely monitoring data, optimizing resource utilization, and increasing productivity.

Keywords: Data ingestion, Fuzzy logic, IoT, Precision farming, Thing speak.

INTRODUCTION

Today's era has witnessed rapid urbanization and industrialization thanks to ongoing technological evolution and continuous advancements, leading to notable

* **Corresponding author Sudheer Mangalampalli:** Department of CSE, Manipal Institute of Technology, Manipal Academy of Higher Education, Manipal, India, E-mail: ms.sudheer@manipal.edu

Suneeta Satpathy, Sachi Nandan Mohanty & Subhendu Kumar Pani (Eds.)

emerging technologies. India has made progress in practically every industry, but has fallen behind in agriculture, as it still relies on antiquated conventional practices. Agriculture is crucial to a country's development, as it is a significant contributor to the country's GDP and ensures food security for a massive population. Implementing new technologies has the potential to revolutionize farming practices. In traditional agriculture, farmers rely on manual labor for crop and livestock management, often resulting in ineffective resource utilization. Precision farming offers hope for overcoming agricultural challenges. It uses GPS, sensors, and prominent computing technologies like the cloud, IoT, and Fuzzy Logic to monitor crop health remotely and provide precise nutrients and water to reduce waste and increase yields. The Internet of Things (IoT) integrates tangible items into the Internet Protocol (IP) framework, enabling them to connect to the Internet and communicate with other systems and devices. These interconnected devices make intelligent decisions based on gathered data. Domains such as neural networks, AI, and ML play a significant role. The output is automated based on training, and real-time data can be visualized on the cloud from anywhere. The proposed precision farming system is designed to collect data from sensors and transmit it to an Arduino, which in turn sends this large spectrum of data to an IoT-based Cloud platform, ThingSpeak, where the data can be viewed in the form of graphs. This platform comes with integrated MATLAB Analysis and Visualization Apps, where we use the Fuzzy Logic Toolbox to regulate a controlled machine output according to multiple inputs.

LITERATURE REVIEW

Various references have been considered and reviewed to identify existing systems and determine their applications, architectures, and flaws. Xu *et al.* review the introduction and implementation of IoT in agriculture, reporting existing problems and future trends [1]. It focuses on the evolution of communication technologies in agricultural IoT, information processing, and environment monitoring, paving the way for sustainable ideas. Chapter 9 of the book titled "Sustainable Digital Technologies for Smart Cities" describes the role of intelligent predictive analytics and big data in farming, comparing Fuzzy Logic to binary logic and justifying this with performance metrics [2]. A study focuses on using Fuzzy Logic for a controlled water pumping mechanism, but could have used additional sensors, like PH, air quality, and light intensity, to achieve more efficient results [3]. Kagalkar addressed in his study the issue of excess water utilization, but proposes manual handling of the water pump [4]. Other studies suggest frameworks for water pumping based solely on soil moisture, which may yield imprecise outcomes under fluctuating humidity and temperature [5, 6]. Astudy discusses the early implementation of IoT in agriculture, with a focus on data collection and cloud visualization [7]. Pradana *et al.* proposed an irrigation control system using the Tsukamoto Fuzzy Algorithm, which is less accurate than the Mamdani FIS due to a high MAPE value [8, 9]. Katthineni and Reddy presented an architecture for motor control *via* relay using IoT, Cloud, and Machine Learning; however, it relies on predicting using datasets in arrays, which can become complex with large data spectra [10]. Kaswar *et al.* specify a hydroponic system for adaptive control of nutrition, but the models could be more interpretable [11]. A study by Abioye *et al.* outlines a monitoring system and data-driven modeling using drip irrigation, which is not cost-effective and can lead to clogging issues [12]. Benyezza *et al.* propose an innovative greenhouse control platform with a Human-Machine Interface on a Node-RED server for data visualization, but it faces complexity issues in debugging and lacks architectural benefits [13]. Pierre *et al.* propose an irrigation system with low energy consumption [14]. It suggests the inclusion of solar panels, which satisfies the aim, and also includes an XBee module as a communication technology. Xbee, although it uses low power to operate and has a flexible network topology, its transmission distance ranges only from 10 to 100 m, which is not an efficient approach considering the development of Wireless WAN, which has a very long transmission distance.

PROPOSED ARCHITECTURE

Control System

The project comprises functional parts, including a motor/water pump, a GSM/WiFi module, an LCD, a solar panel, and various sensors, such as soil moisture, air humidity, air quality, temperature, and light intensity sensors, all connected to an Arduino Uno, an open-source electronics prototyping platform. This microcontroller can receive input from sensors and initiate actions, like water pumping. The GSM/WiFi module, which uses a SIM card for connection *via* an RS-232 interface, sends and receives messages through GPRS using AT (Attention) Commands. The LCD displays sensor values and is placed near the microcontroller. Solar panels power the project, utilizing renewable energy for long-term efficiency and low operational costs. The soil moisture sensor measures soil water content in percentage volume. The DHT11 sensor records environmental temperature and humidity. The Light Dependent Resistor (LDR) sensor measures light levels around the crop, with resistance decreasing as light intensity increases. The MQ2 air quality sensor evaluates air quality based on chemo-resistive conductivity, identifying gases, like methane and carbon dioxide, and measuring them in ppm (parts per million), with higher ppm values indicating poorer air quality.

Fuzzy Logic

It is an algorithm that allows sophisticated decision tree processing and improved rules-based programming integration. It has a range of real values between 0 and 1, *i.e.*, it can have partial truth values. This makes real-life scenarios seem more plausible. Agriculture is a sector where variations in data are collected over time because of various unprecedented situations. Fuzzy logic is used to determine logical relationships from this extensive and diversified data. It consists of three steps: fuzzification, rule inference, and defuzzification. Mamdani type 1 is the Fuzzy Inference System implemented in this instance because of its widespread acknowledgment and more interpretable rule base. Fig. (**1**) shows the system architecture, and Fig. (**2**) shows the Trapezoidal membership graph.

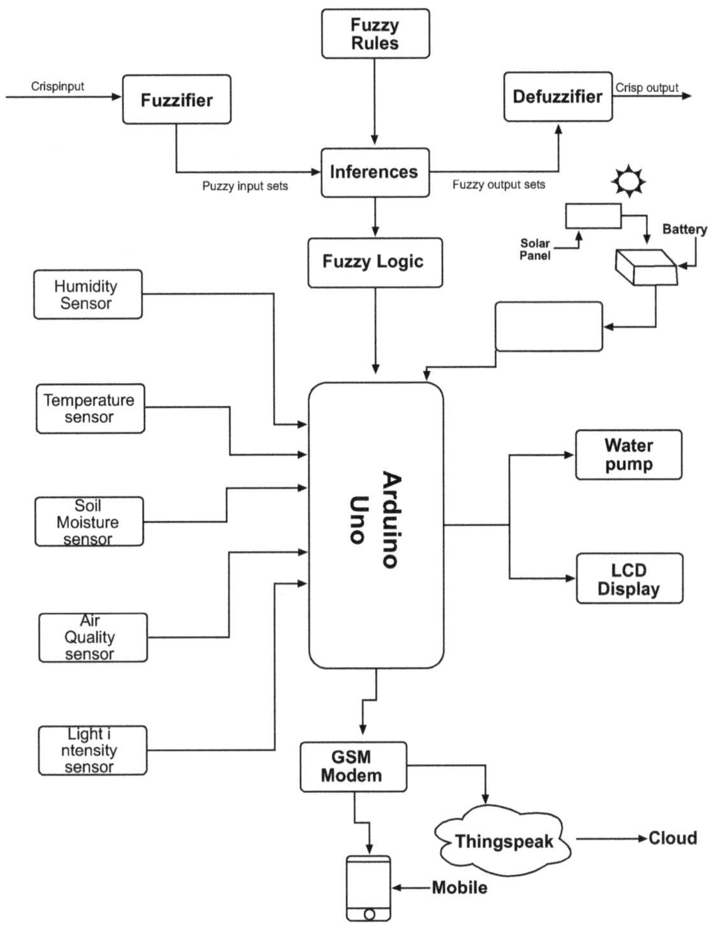

Fig. (1). System architecture

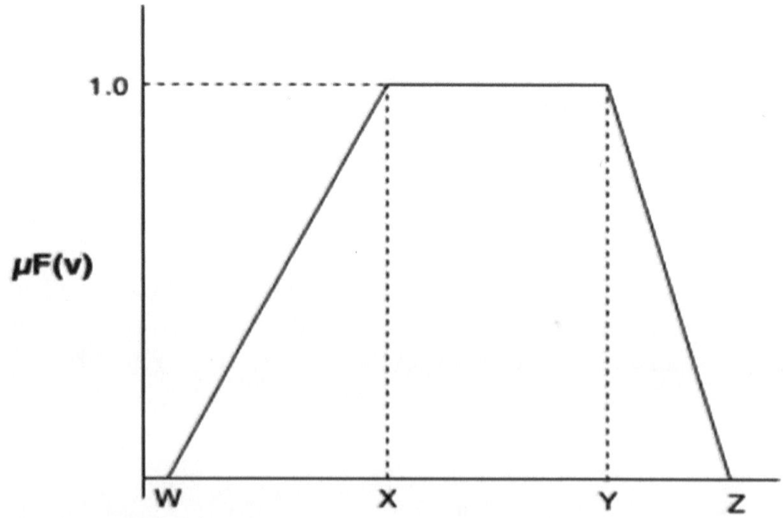

Fig. (2). Trapezoidal membership graph.

METHODOLOGY

Fuzzy Logic involves defining linguistic variables and terms, followed by constructing membership functions for them, and then generating rules logically applicable. Rule evaluation is carried out through fuzzy set operations. Min and Max operations are employed for OR and AND. Combining all these results, we can form a final result. Finally, defuzzification is done according to the membership function for the output variable.

Crisp input values are transformed into fuzzy values in the first stage, called fuzzification. A fuzzy set determines values between 0 and 1. The membership function denotes the extent of affiliation with the fuzzy set, represented by μF(v), where F is a fuzzy set. The higher the value in [0,1], the more the degree of belonging [15].

μF(v): V à [0, 1] where μF(v) = 1 if v is totally in the fuzzy set or μF(v) = 0 if v is not in the fuzzy set and 0 < μF(v) < 1 if v is partly in the fuzzy set.

Trapezium-shaped membership function plots generally have four parameters. Let w, x, y, and z represent the x coordinates of the membership function (1) [16]: Then,

$$\mu trapezoid = max\ (min\ (a\text{-}w/x\text{-}w,\ 1,\ z\text{-}a/z\text{-}y),0) \quad \mathbf{(1)}$$

The project suggests five sensor variables accompanied by a specific membership set. The creation of fuzzy sets is depicted in Table **1**. Fuzzy logic is employed to evaluate inputs with multiple values, generating a unified multi-valued outcome.

Table 1. Fuzzy sets.

$\mu_{AH}^{Low}(v)$=0, ∀v ≥ 85%	$\mu_{AH}^{Low}(v)$=1, ∀v ≤ 60%
$\mu_{AH}^{High}(v)$=0, ∀v ≥ 70%	$\mu_{AH}^{High}(v)$=1, ∀v ≥ 85%
$\mu_{AT}^{Cool}(v)$=0, ∀v ≥ 30 °C	$\mu_{AT}^{Cool}(v)$=1, ∀v ≤ 10 °C
$\mu_{AT}^{Hot}(v)$=0, ∀v ≤ 10 °C	$\mu_{AT}^{Hot}(v)$=1, ∀v ≥ 30 °C
$\mu_{SM}^{Dry}(v)$=0, ∀v ≥ 35	$\mu_{SM}^{Dry}(v)$=1, ∀v ≤ 20
$\mu_{SM}^{Moderate}(v)$=0, ∀v ≤ 20 & ∀v ≤ 45	$\mu_{SM}^{Moderate}(v)$=1, 35 ≤ v ≤ 45
$\mu_{SM}^{Wet}(v)$=0, ∀v ≤ 35	$\mu_{SM}^{Wet}(v)$=1, ∀v ≥ 50
$\mu_{LI}^{Dim}(v)$=0, ∀v ≥ 450	$\mu_{LI}^{Dim}(v)$=1, 200 ≤ v ≤ 300
$\mu_{LI}^{Normal}(v)$=0, ∀v ≤ 600	$\mu_{LI}^{Normal}(v)$=1, 400 ≤ v ≤ 500
$\mu_{LI}^{Bright}(v)$=0, ∀v ≤ 550	$\mu_{LI}^{Bright}(v)$=1, ∀v ≥ 800
$\mu_{AO}^{Low}(v)$=0, ∀v ≥ 800	$\mu_{AO}^{Low}(v)$=1, ∀v ≤ 700
$\mu_{AO}^{High}(v)$=0, ∀v ≥ 600	$\mu_{AO}^{High}(v)$=1, ∀v ≥ 1000
$\mu_{WT}^{VeryShort}(v)$=0, ∀v ≥ 8	$\mu_{WT}^{VeryShort}(v)$=1, ∀v ≤ 2
$\mu_{WT}^{Short}(v)$=0, ∀v ≤ 3 & ≥16	$\mu_{WT}^{Short}(v)$=1, 8 ≤ v ≤ 12
$\mu_{WT}^{Average}(v)$=0, ∀v ≤ 12 & ≥35	$\mu_{WT}^{Average}(v)$=1, 18 ≤ v ≤ 31
$\mu_{WT}^{Long}(v)$=0, ∀v ≤ 30 & ≥55	$\mu_{WT}^{Long}(v)$=1, 35 ≤ v ≤ 45
$\mu_{WT}^{VeryLong}(v)$=0, ∀v ≤ 45	$\mu_{WT}^{VeryLong}(v)$=1, ∀v ≥ 55

The values specified on the left side of the table do not pertain to the membership sets, while the values provided on the right side of the table represent valid values that belong to the Fuzzy Set. The table employs certain abbreviations: “AH” refers to Air Humidity, “AT” indicates Air Temperature, “SM” represents Soil Moisture, “LI” stands for Light Intensity, “AQ” denotes Air Quality, and “WT” signifies Watering Time.

The MATLAB fuzzy inference system offers users a wide range of options to generate rules and to analyse their performance. One can create own rules by adding all the necessary parameters or generate all possible rules for input and output combinations.

Defuzzification (the procedure of transforming a fuzzified output into a singular crisp value) is done using the centroid method. Within Mamdani Fuzzy Inference Systems (FIS), rules are formulated employing linguistic variables for both the premise and conclusion. The formula applied for the centroid defuzzification method within the Mamdani FIS is as indicated below (2):

$$A = \int \mu F(x).\, xdx \, / \int \mu F(x)dx \qquad \textbf{(2)}$$

Fig. (**3**) represents the Fuzzy rules used in this research. The range for input and output variables is as follows: Input sensor variables range: Humidity sensor: 50-100%, Soil Moisture sensor: 0-60%, Air Temperature sensor: 5-50°C, Light Intensity sensor: 100-1000 photocell reading, Air Quality sensor: 500-1200 ppm, Output variable range: Water Pumping Time (WPT): 0-60s. Fig. (**4**) shows the editor for FIS, which displays graphs for both input and output variables along with their membership functions. The FIS (Fuzzy Inference System) used here is a Mamdani Type 1, notably with 5 inputs and 1 output. Properties of the FIS are: And Method: “min”, Or Method: “max”, Implication Method: “min”, Aggregation Method: “max” and Defuzzification Method: “centroid”. On the x-axis of each plot, we can observe the previously defined range for each input and output variable. The y-axis represents the degree of membership. A degree of 0 indicates that the value does not belong to the set. As the degree of membership increases towards 1, values in between progressively align more closely with the set’s characteristics. When the degree of membership reaches 1, it signifies that the value lies entirely within the set’s boundaries.

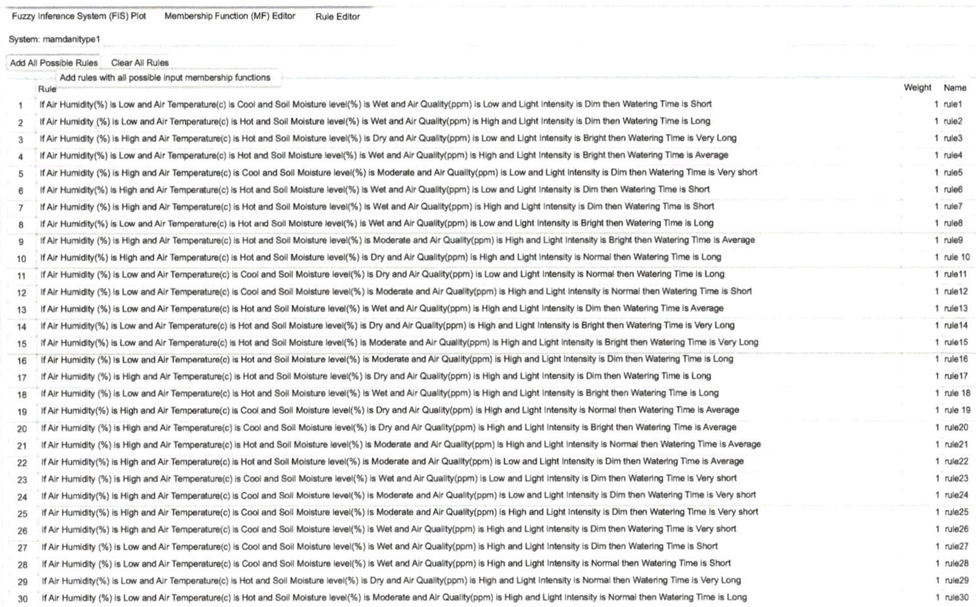

Fig. (3). Fuzzy rules.

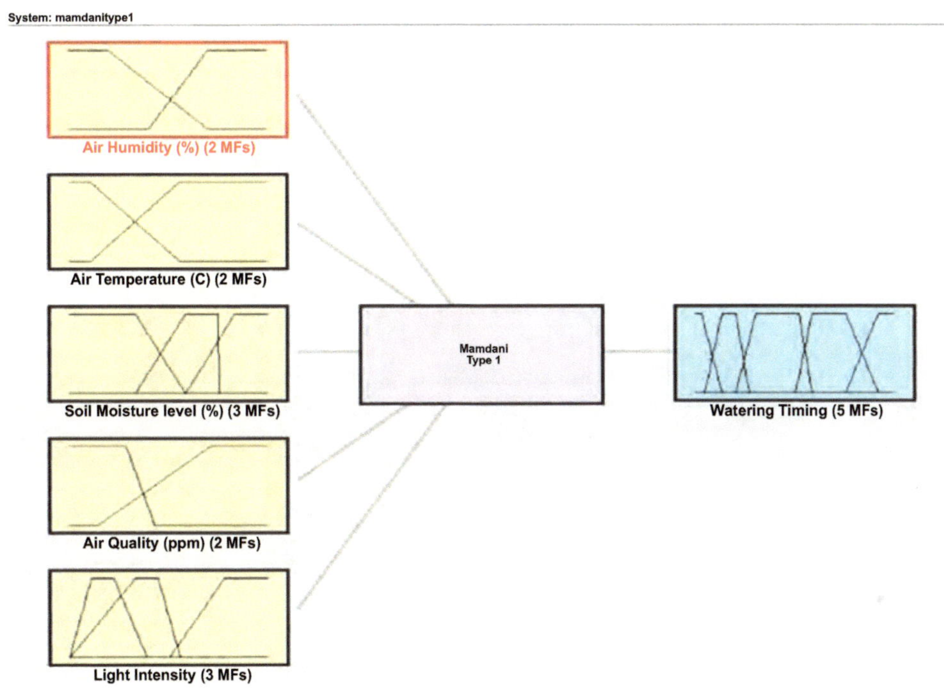

Fig. (4). Editor for the fuzzy inference system.

RESULTS AND DISCUSSION

The system displays field charts for soil moisture, humidity, temperature, light intensity, and air quality over time using ThingSpeak.com. The channel features five charts, each representing data from different sensors. The x-axis shows time, which can be adjusted to display either date, time, or months, while the y-axis shows the recorded values. Sensors capture data at set intervals, allowing farmers to analyze daily data, especially during crop growth. This setup also enables farmers to review data from previous years for informed planting decisions. ThingSpeak securely stores data in the cloud, making it accessible at any time. Its simplicity and user-friendly interface ensure quick and easy adoption by the farming community.

Fig. (**5**) suggests that with an average air humidity, an average temperature, and soil moisture ranging between moderate and wet, the output watering time is approximately 30.7 seconds. It visually represents the correlation between input and output variables using 3D diagrams, analysing the relationship between sensor readings and watering time fluctuations. Fig. (**6**) shows that watering time increases with rising light intensity and decreasing air humidity.

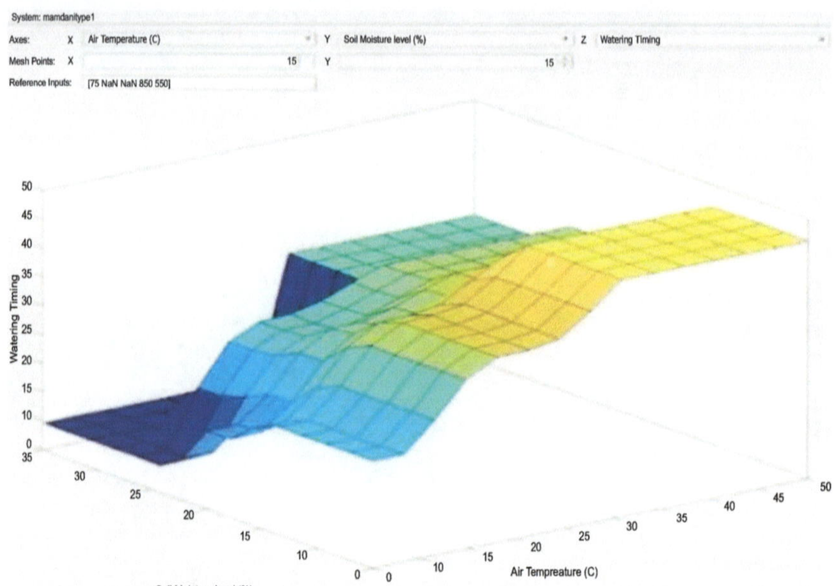

Fig. (5). Surface Viewer (temp and soil moisture *vs* Water pumping time).

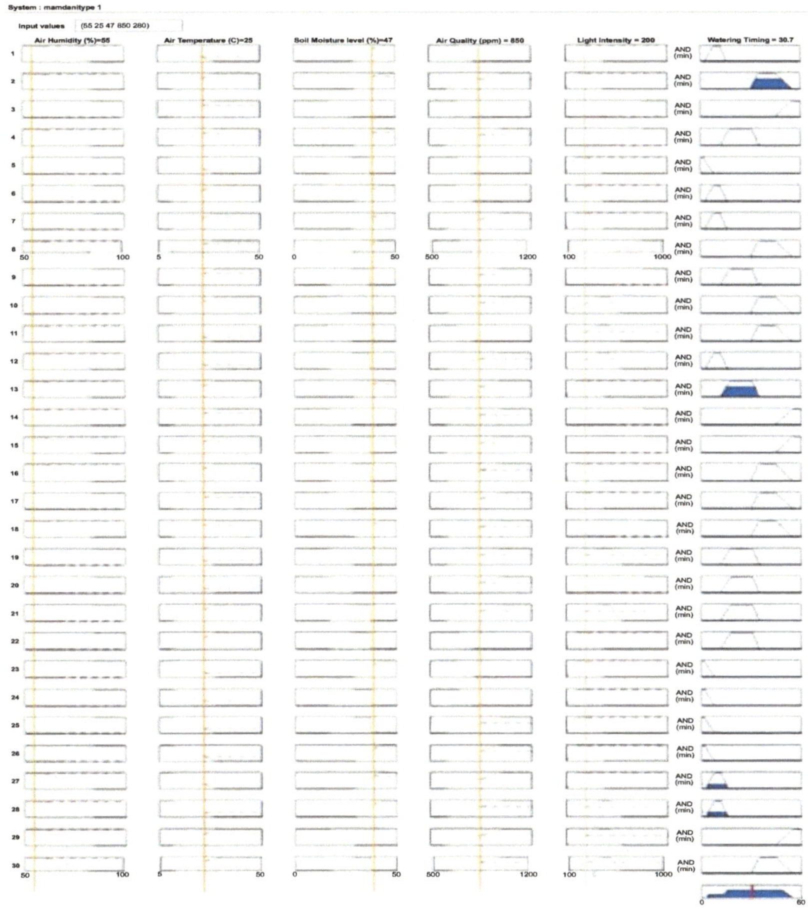

Fig. (6). Ruler viewer of the Fuzzy System [55%, 25 °C, 47%, 850 ppm, 280 photocell reading].

CONCLUSION

Automation has been successful in every field where it has been implemented. Agriculture is one such sector that cannot be disregarded as it is vital for sustenance. Therefore, the implementation of automation in this industry can produce better results in terms of productivity compared to those produced using traditional methods, while also reducing workload and physical labour. This project has worked on five sensors and training data, using fuzzy logic, which has scope for improvement in the coming years. The primary focus is on soil profile analysis, a critical component of agricultural practices, as it offers insightful data on the traits and qualities of the soil in a particular region. So, in future implementations, we can analyze the soil layers or horizons and work on providing nutrients/ supplements that are in deficit to the soil during each stage of plant growth. The water level in the tank is one issue that develops even though the output is efficient in terms of watering. To avoid instances like turning on the pump when there is no more water in the tank, we can add this problem to the SMS generator area to inform the farmer when the water level in the tank falls below a specific mark. The pH of soil is another concern, as few crops are pH-sensitive. We can deploy a pH sensor that can record the pH value of the soil. A larger dataset can be efficient in gaining more accurate findings after the training and testing phase. Another implementation could be analysing the environmental conditions over a period of time (during all seasons) and then suggesting to farmers which kind of crop would produce high yields and earn them maximum profits.

ACKNOWLEDGEMENTS

The authors acknowledged the support of the Department of CSE, Manipal Institute of Technology, Bengaluru, Manipal Academy of Higher Education, Manipal, India.

REFERENCES

[1] J. Xu, B. Gu, and G. Tian, "Review of agricultural IoT technology", *Artificial Intelligence in Agriculture,* vol. 6, pp. 10-22, 2022. [http://dx.doi.org/10.1016/j.aiia.2022.01.001]

[2] S. Sangeetha and N. Suganthi, “Intelligent predictive analytics in farming using IoT in big data,” in Sustainable Digital Technologies for Smart Cities, CRC Press, 2023, pp. 87–98. [http://dx.doi.org/10.1201/9781003307716-9]

[3] N. Abdullah, N.A.B. Durani, M.F.B. Shari, K.S. Siong, V.K.W. Hau, W.N. Siong, and I.K.A. Ahmad, "Towards smart agriculture monitoring using fuzzy systems", *IEEE Access,* vol. 9, pp. 4097-4111, 2021. [http://dx.doi.org/10.1109/ACCESS.2020.3041597]

[4] A. Kagalkar, "Smart irrigation system", *Int. J. Engine Res.,* vol. 6, no. 5, 2017.

[5] C.A. Ihendinihu, "Inyama k. Development of smart irrigation system based on internet of things (IoT", *Development,* vol. 9, no. 5, 2023.

[6] B.A. Tapakire, and M.M. Patil, "Iot based smart agriculture using thingspeak", *Int. J. Eng. Res. Technol. (Ahmedabad),* vol. 8, no. 185, pp. 10-7577, 2019.

[7] H.C. Punjabi, S. Agarwal, V. Khithani, V. Muddaliar, and M. Vasmatkar, "Smart farming using IoT", *International Journal of Electronics and Communication Engineering and Technology.,* vol. 8, no. 1, pp. 58-66, 2017.

[8] F.G. Pradana, R. Sarno, and S. Triarjo, "Development of Smart Farming Control System based on Tsukamoto Fuzzy Algorithm", *2023 International Conference on Computer Science, Information Technology and Engineering (ICCoSITE),* pp. 348-353, 2023. [http://dx.doi.org/10.1109/ICCoSITE57641.2023.10127754]

[9] E. Sonalitha, B. Nurdewanto, S. Ratih, N. R. Sari, A. B. Setiawan, and P. Tutuko, “Comparative analysis of Tsukamoto and Mamdani fuzzy inference system on market matching to determine the number of exports for MSMEs,” in Proc. 2018 Electrical Power, Electronics, Communications, Controls and Informatics Seminar (EECCIS), Oct. 2018, pp. 440–445. [http://dx.doi.org/10.1109/EECCIS.2018.8692989]

[10] K. Shilpa and V. Ganesh Babu, “A Study on Smart Agriculture Monitoring Systems Using IoT and Machine Learning Technology,” International Journal of Science, Engineering and Technology, vol. 7, no. 2, Mar–Apr 2019.

[11] A.B. Kaswar, R.D. Mahande, and J.D. Malago, "a New Model for Hydroponic Lettuce Nutrition Adaptive Control System Based on Fuzzy Logic Sugeno Method Using Esp32", *Jurnal Teknik Informatika (Jutif),* vol. 4, no. 2, pp. 391-400, 2023. [http://dx.doi.org/10.52436/1.jutif.2023.4.2.626]

[12] E.A. Abioye, M.S.Z. Abidin, M.S.A. Mahmud, S. Buyamin, M.K.I. AbdRahman, A.O. Otuoze, M.S.A. Ramli, and O.D. Ijike, "IoT-based monitoring and data-driven modelling of drip irrigation system for mustard leaf cultivation experiment", *Inf. Process. Agric.,* vol. 8, no. 2, pp. 270-283, 2021. [http://dx.doi.org/10.1016/j.inpa.2020.05.004]

[13] H. Benyezza, M. Bouhedda, R. Kara, and S. Rebouh, "Smart platform based on IoT and WSN for monitoring and control of a greenhouse in the context of precision agriculture", *Internet of Things,* vol. 23, p. 100830, 2023. [http://dx.doi.org/10.1016/j.iot.2023.100830]

[14] N.J. Pierre, B. Sefu, S. Venuste, M.J. D’Amour, K. Daniel, N.J. Pierre, K.J. Bosco, and H. Felix, "Smart crops irrigation system with low energy consumption", *Academic Society for Appropriate Technology,* vol. 9, no. 1, pp. 9-19, 2023. [http://dx.doi.org/10.37675/jat.2022.00192]

[15] S. Gupta, P.K. Biswas, B. Aljafari, S.B. Thanikanti, and S.K. Das, "Modelling, simulation and performance comparison of different membership functions based fuzzy logic control for an active magnetic bearing system", *J. Eng.,* vol. 2023, no. 2, p. e12229, 2023.

[16] Samanta D. *Fuzzy membership function formulation and parameterization.* İnternet Adresi: Http://Www. Nid. İitkgp. Ernet. İn/Dsamanta/Courses/Archive/Sca/Archives. 2018.

CHAPTER 5

Enhancing Feedforward Neural Network Optimizer Variants to Strengthen Aquaponics Fish Growth Predictions

Safiya Begam[1], **W. Aisha Banu**[1,*] and **Sharon Priya**[1]

[1]*B. S. Abdur Rahman Crescent Institute of Science and Technology, Chennai, Tamil Nadu, India*

Abstract: For training a deep learning model, one has to get the parameters necessary to satisfy an objective function. In general, the objective is to minimize the loss encountered throughout the learning process. A type of supervised learning involves feeding data samples and their matching outputs to a model. A framework works to get the generated output nearer to the desired result by comparing its results to the planned outcome and taking into account any variations. Optimization algorithms have been utilized to do this. Multiple cycles of optimization are carried out till the process is completed in order to increase the accuracy of the model. Numerous optimization techniques have been developed to address the challenges associated with the learning process. The techniques addressed involve Adadelta, Adagrad, Adam, and stochastic gradient descent. With Adam and Adagrad, Pond1 has a desirable training result of 0.96 at epoch 200, and Pond2 has an optimal training result of 0.96 at epoch 250. For Ponds 1 and 2, the best Adam test results are 0.9936 and 0.9975, respectively. Except for AdaGrad, which yielded the lowest scores, the results show that Adam-based optimizers gave the best results. Depending on the surroundings, an aquaponics system with a well-optimized FNN may help make more accurate predictions of fish development.

Keywords: Aquaponics, Feedforward neural network, Optimizer variants, Predictive analytics.

INTRODUCTION

Combining hydroponics and aquaculture, aquaponics is a sustainable farming method that offers enhanced resource efficiency, variation, and less negative environmental effects. Most beneficial are urban regions with little arable land and water-stressed places. Aquaponics promotes local food production, reduces reliance on artificial fertilizers and pesticides, and enables year-round production

* **Corresponding author W. Aisha Banu:** B. S. Abdur Rahman Crescent Institute of Science and Technology, Chennai, Tamil Nadu, India; E-mail: aisha@crescent.education

Suneeta Satpathy, Sachi Nandan Mohanty & Subhendu Kumar Pani (Eds.)

of fresh, healthy food. It also reduces the carbon footprint associated with long-distance travel and offers an opportunity for experiential learning [1]. An irrigation system, known as an aquaponics system, combines fish and plants. In this cyclical structure, nitrifying bacteria break down fish excrement, which is then eventually absorbed by the plants as food. Because these aquaponics systems can be installed practically anywhere, they are ideal for urban farming. The study "Food, Fish Growth, and Fish Yield Relationships" looks into the relationships between pond fish productivity, fish development, and food availability. It examines dietary choices, feed mix, and nutritional characteristics, and how these factors affect fish development. Research is essential for managing fisheries and aquaculture [2].

Getting concerned about the water, such systems require a significant initial investment, even though they should consume less energy, water, and fertilizers than conventional agricultural systems. This issue provides important insights for researchers studying fish physiology and aquaculture by examining the endocrinological aspects of fish growth and reproduction. It focuses on the mechanisms of action and effects of different hormones on physiological processes [3]. Knowledge of fish growth is essential for maintaining a healthy aquaponics environment, as overfishing and the biological and atmospheric content of the substrate—which is influenced by pollution among other things—both negatively affect fish growth. Fish populations need to be regularly checked for the aquaponics system to be successful. Diversifying the systems is equally important as raising productivity and/or species diversity to support this expansion. In aquaponics, the total efficiency of the system depends on the fish's ability to develop normally. The entire productivity and sustainability of the setup are directly affected by it. Healthy aquatic environments ensure nutrient-rich waste, which is essential for plant growth, and stable fish growth is a sign of this. This mutually beneficial relationship, in which fish provide vital nutrients to plants and gain from a hygienic and cleansed drinking environment, is the foundation of aquaponics management. Ensuring the health of the fish population requires monitoring and controlling parameters related to fish development, such as temperature, water quality, and feeding. A sustainable supply of protein for human consumption is also provided by fish that grow efficiently, which benefits the system's nutrition cycle [4]. The goal of sustainable development is to strike a balance between human and ecological well-being.

An ecosystem approach to the fishing industry has recently gained recognition as a significant area of study. The article's structure is best described as follows: Section II covers the pertinent research on optimization approaches. The datasets and optimization techniques used in this work are covered in Section III. The algorithms Adam, Adagrad, Adadelta, and stochastic gradient descent are covered. The IoT-based Aquaponics datasets for Ponds 1 and 2 are the ones being discussed. In Section IV, the results of training the model with the available optimization techniques on pertinent datasets are compared and elaborated upon. In Section V, the paper is summarized, and future directions are discussed.

LITERATURE REVIEW

The taxonomy presented in the article facilitates a clear understanding of the environment by categorizing different approaches within the field of deep learning. However, the work needs to be improved because it does not adequately address the difficulties in implementing deep learning technologies, new approaches, and their interactions with reinforcement learning, as well as their ethical and societal ramifications.

An extensive overview of deep learning methods, taxonomy, applications, and future possibilities is given in Sarker's 2021 paper. It categorizes approaches and highlights their importance. Future directions and current research trends are also covered in the paper [5].

In 2021, Naseri conducted a study to examine the impact of varying parameters on both traditional and deep neural networks. The study highlights the importance of comprehending how these networks react to modifications in activation functions, learning rates, and network topology for effective task design and fine-tuning [6].

In order to enhance training and convergence, this paper provides a revised version of the Adam optimizer for deep neural networks (DNNs). The enhancement addresses flaws in the first methodology, potentially enhancing the efficacy and efficiency of deep neural network training. Researchers and practitioners interested in machine learning and neural network optimization will find this study to be of interest [7].

This effort aims to study optimizers in scientific computing and deep learning. It discusses several optimization algorithms, including SGD, Adam, and RMSprop, outlining their principles, advantages, and potential applications. Effective neural network training requires an understanding of optimizers, making this study valuable for both practitioners and scholars [8].

The study describes an approach for selecting suitable deep learning algorithms based on specific problem settings or datasets. It utilizes machine learning techniques to analyze the features of the dataset and matches them with the relevant algorithms. Researchers and practitioners interested in enhancing the performance and efficiency of models can benefit from this resource on machine learning and deep learning [9].

Optimization strategies for enhancing the performance of deep neural networks are analyzed in this article. It assesses well-known techniques, such as Adam, Momentum, NAG, and Stochastic Gradient Descent. In order to provide insights for successful and efficient training procedures, the analysis attempts to determine the most effective strategy for training DNNs. For scholars and practitioners alike, this research is essential [10].

This research examines how Convolutional Neural Networks (CNNs) use optimization techniques for binary classification tasks. Using standard optimizers like Adam, RMSprop, and Stochastic Gradient Descent, it analyzes the results to find the most accurate and efficient classification. This study helps researchers and practitioners choose the most effective optimization method. By contrasting less-studied optimization strategies and understanding the influence of optimizer selection on computing resource requirements, it might close a research gap. This could clarify their applicability and possible advantages in certain situations, especially when resources are scarce [11].

The study compares deep learning optimizers empirically and evaluates their performance across a range of tasks. The goal is to comprehend the effects of various optimizers on model accuracy, robustness to noisy data, and convergence speed. This study helps academics and practitioners select the best optimization approach for the particular datasets and applications they are working with. Examining optimizers' performance on distinct or specialized architectures and in certain scenarios, like with insufficient training data or unbalanced class distributions, fills a research vacuum in deep learning optimization. Filling these gaps could help practitioners working on practical applications and advance our understanding of optimizer behavior in a wide range of challenging and diverse deep learning scenarios [12].

Analyzing and contrasting optimization techniques for deep learning model training in image recognition tasks is the aim of this study. To offer advice on choosing the best optimizer, it aims to understand how various optimizers affect training convergence time, final classification accuracy, and robustness to image fluctuations. A knowledge gap exists regarding the performance of optimizers under specific conditions, such as picture noise, resolution, or deformities, in research on deep learning-based image identification optimizers. This could improve their performance in a range of applications beyond photo identification tests by revealing how they adapt to real-world situations and can be generalized across different deep learning architectures [13].

This study provides a novel approach to improving network security *via* the detection of intrusions through the use of deep learning and optimization techniques. The strategy integrates artificial intelligence to enhance the accuracy and efficiency of the intrusion detection system (IDS). Deep neural networks are used to extract features and identify patterns, while optimization algorithms are used to adjust the model and improve its performance. According to the study, this hybrid approach offers a strong response to the rising problem of cybersecurity in multifaceted network environments by significantly enhancing the ability to detect and neutralize attacks against networks in real-time [14].

An innovative method for guaranteeing the correctness and dependability of multi-class feed-forward neural networks is presented in this study. The technique is centered on quantitatively validating the network's behavior in a range of input contexts, which aids in detecting possible mistakes or discrepancies in categorization assignments. The method offers a mechanism to evaluate and confirm the model's performance by methodically examining the network's answers, particularly in crucial applications where accuracy is crucial. By using this verification technique, neural networks become more resilient and perform as expected on a variety of intricate datasets [15].

The forward-backward algorithms are widely used in signal processing, machine learning, and image reconstruction. The suggested method uses deep neural networks to forecast and accelerate the iterative steps in these algorithms. The strategy drastically reduces calculation time without sacrificing or even raising the accuracy of the results by utilizing deep learning to guide the optimization process. Large-scale issues that require a significant amount of computing power benefit greatly from this acceleration strategy [16].

The importance of hyperparameter optimization in simulating the intricate, nonlinear patterns of wind power generation is emphasized in the study. To optimize model performance, the researchers use genetic algorithms, Bayesian optimization, and other cutting-edge techniques to fine-tune the model's parameters. The findings show that these cutting-edge optimization strategies greatly enhance the deep learning models' predictive capacities, resulting in more precise and trustworthy wind power projections—an essential component of incorporating renewable energy sources into the electrical grid [17].

This article investigates the effects of various weight initialization strategies, especially when combined with diverse activation functions, on the performance of feedforward neural networks. This paper investigates the impact of weight initialization decision affects the rate of convergence, stability, and total accuracy of neural networks during training. It compares and assesses the efficacy of standard initialization techniques, including Xavier, He, and uniform initialization, in conjunction with activation functions like as sigmoid, tanh, and ReLU. The results emphasize the importance of selecting the appropriate weight initialization procedures to enhance model performance, reduce training times, and prevent issues such as vanishing or exploding gradients—all of which contribute to neural network architectures that are ultimately more efficient [18].

This study introduces a cutting-edge approach to enhancing deep learning models through a novel hybrid method that combines the Bees Algorithm with Bayesian optimization techniques. This hybrid Bees Bayesian approach is applied to Convolutional Neural Networks (CNNs) to optimize their parameters more effectively. The integration of the Bees Algorithm, known for its ability to explore and exploit search spaces, with Bayesian optimization, which efficiently hones in on optimal parameter values, leads to improved model accuracy and efficiency. The study demonstrates that this novel hybrid approach significantly outperforms traditional optimization methods, making it particularly valuable for complex tasks where CNNs are deployed, such as image recognition and classification [19].

A novel approach to heart disease classification that uses a deep learning model improved using Crow Intelligence Optimization (CIO). This method uses the CIO algorithm's special ability to solve a problem—which resembles the clever way crows find the best answers—to adjust the parameters of the deep learning model. Because it can distinguish between various cardiac problems, the final model is more effective and accurate in detecting heart illness. The study shows that the performance of heart disease classification is significantly enhanced by this optimization-based method, providing a potent instrument for early diagnosis and improved patient outcomes in medical applications [20].

A comprehensive examination of the several methods used to optimize deep neural networks, including the hyperparameters, which are essential for improving model performance. It covers more sophisticated methods like reinforcement learning, genetic algorithms, and Bayesian optimization in addition to more conventional strategies like random and grid search. The study emphasizes the importance of tuning hyperparameters effectively to achieve the optimal model accuracy and efficiency. It also covers how automated machine learning (AutoML) can help make the hyperparameter optimization process easier to understand and more scalable for intricate applications [21].

FNN, OPTIMIZATION STRATEGIES AND DATASETS

A basic kind of artificial neural network is called a feedforward neural network (FNN), where data travels in a single route from nodes that supply inputs to the final nodes through hidden layers that can customized as shown in Fig. (**1**). Each unit adds non-linearity by applying an activation function to the weighted sum of its inputs and having a weighted connection. The network processes input data during training, and a choice of loss functions is used to compare the output to the real targets. Gradients are then calculated and used to iteratively change the network's parameters, minimizing the loss function throughout the backpropagation process. The network can classify new data or make predictions after it has been trained. Because of their versatility, FNNs are frequently used for tasks including pattern recognition, regression, and classification, among others [22].

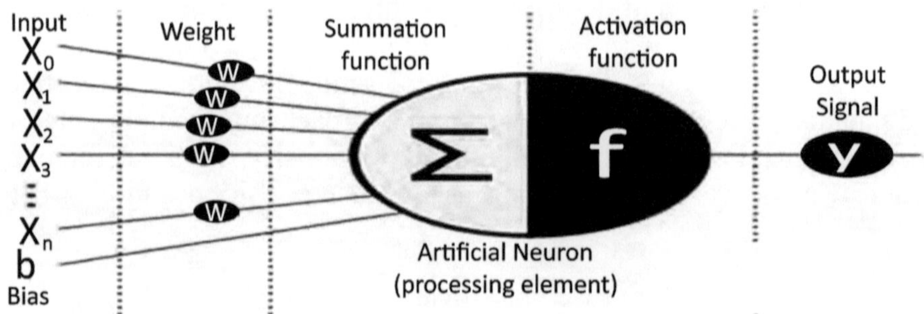

Fig. (1). Feed forward neural network model (Reference:https://doi.org/10.1007/s42979-021-00815-1).

Neural network training involves the use of a neural network optimizer, which is crucial for effective training. Its job is to minimize the error or loss function by adjusting the internal parameters of the network (weights and biases) throughout the learning phase. In order to determine the magnitude and direction of change for the parameters, the optimizer uses a number of optimization approaches. In deep learning, optimizers are algorithms that change the model's parameters to minimize a loss function during training. By iteratively updating their weights and biases, neural networks learn from data. Among the well-known optimizers are Adam, RMSprop, and Stochastic Gradient Descent (SGD). Each optimizer has a unique set of training rates, momentum, and revision criteria to determine the best model variables [23]. Optimization methods are processes that are optimized to assist a deep learning model perform better. The accuracy and speed at which deep neural networks are trained are greatly impacted by these optimization methods, also known as optimizers. Through the use of deep learning and optimization approaches, research attempts to improve service quality in a variety of applications. In order to improve performance in real-time applications, online gaming, and multimedia streaming, neural networks must be trained to predict quality of service (QoS) parameters [24].

SGD

The chance upon which the technique is based is referred to be stochastic. With stochastic gradient descent (SGD), data batches are randomly chosen for each iteration rather than the entire dataset. The cost function minimizes (EPOCH) following each iteration. It modifies the objective function (cost function) at regular intervals with a high degree of fluctuation.

AdaGrad

Here, the adaptive gradient optimization algorithm is employed, and the updated parameter values are determined mostly by the program's learning rate. This optimizer uses a different learning rate for each iteration (EPOCH) instead of an identical training rate for all parameters, despite stochastic gradient descent. It thus executes larger changes (higher learning rates) for weights corresponding to low-frequency features and smaller updates (fewer training levels) for values that correspond to frequent appearances of data, leading to increased performance with higher accuracy. Adagrad excels at handling sparse data. The alpha at time t is thus calculated first at each iteration, and as the loops rise, the value of t grows, guaranteeing that the alpha t starts to ascend. From point (1) to (4), it is visible.

$$\alpha_j = \sum_{k=1}^{j} \frac{\partial L}{\partial w_k}^2 \tag{1}$$

$$\gamma_j = \frac{\gamma}{\sqrt{\alpha_j + \epsilon}} \tag{2}$$

$$\alpha_j \uparrow \approx \gamma \downarrow \tag{3}$$

$$w_j = w_{j-1} - \gamma \frac{\partial L}{\partial w_{j-1}} \tag{4}$$

AdaDelta

This improves upon the aggressive inclination of the Adaptive Gradient optimizer to decrease the learning rate infinitely. The learning rate can be reduced to a minimal value by expressing the sum of gradients as a decreasing weighted average of all past squared gradients (weighted averages), as opposed to using the preceding squared gradients. For every cycle, the weighted average is calculated first. The restricting term (gamma = 0.95) in Equations (5) and (6) helps us avoid the Vanishing Gradient problem.

$$w_{avg} = r w_{avg_{(j-1)}} + (1 - r)\left(\frac{\partial L}{\partial W}^2\right) \tag{5}$$

$$\gamma_j = \frac{\gamma}{\sqrt{w_{avg} + \epsilon}} \tag{6}$$

Adam

This approach, called Adaptive Moment Estimation, calculates adaptive learning rates for every parameter at every iteration. A combination of RMSprop and Gradient Descent with Momentum is used to calculate the parameter values. Combining the benefits of RMSProp and Gradient with Momentum, including their ability to calculate large amounts of data and variables efficiently, their suitability for irregular objectives, and their low storage requirements. This uses the same adaptive learning rate algorithm and stores the previous squared derivative of loss with respect to weight at time t-1 as a constant weighted average. It has several parameters, such as □, β1, and β2, where Momentum and RMSprop's initial constraining parameters are denoted by β1 and β2. Here, the first moment is denoted by β1, and the second by β2. Compute the first and second moments, which are provided by the following formulas, (7), (8), and (9), in order to update the weights in accordance with a learning rate that adjusts to each iteration t.

$$V_{dw}^{corrected} = \frac{V_{dw}}{\left(1-\beta_1^J\right)} \tag{7}$$

$$S_{dw}^{Corrected} = \frac{S_{dw}}{\left(1-\beta_2{}^t\right)} \tag{8}$$

$$W = w - \gamma \frac{V_{dw}}{\sqrt{S_{dw} + \epsilon}} \tag{9}$$

DATASET

Creating a labelled dataset for an aquaponics fish pond water quality monitoring system would involve labeling data with proper timestamps and collecting information from sensors that detect things like temperature, pH, dissolved oxygen, ammonia levels, nitrate levels, and so forth. In this article, the growth factor for just two ponds is calculated concerning the fish's weight and length.

METHODOLOGY

This method uses a feedforward neural network (FNN) to analyze and forecast further using the calculated growth factor that is obtained from the dataset. First, key input characteristics that are assumed to affect growth are identified, including feed composition, pH levels, water temperature, and nutrient concentrations. These features are then preprocessed to guarantee consistent scaling. Next, the dataset is split up into test, validation, and training sets.

Depending on the complexity of the relationships being modeled, the number of input nodes and hidden layers in the FNN's architecture is chosen. After that, the network is trained on the training set using an optimization technique like SGD or Adam, with the intended output being the previously established growth factor.

When aquaponics pond fish are raised, critical water quality indicators like pH, electrical conductivity (EC), temperature, nitrite levels, and nitrate concentrations are fed into a feedforward neural network (FNN). These metrics are essential indicators of the overall health and condition of the aquaponics environment. The growth condition factor, a crucial metric that represents the well-being and progress of the fish population, is what the FNN is trained to predict. This architecture's use of deep learning optimizers, like Adam or RMSprop, enables the network to comprehend the intricate relationships that exist between these input features and the growth condition factor.

The Feedforward Neural Network (FNN) was trained using various optimization techniques to evaluate the growth factor in the aquaponics system. Stochastic Gradient Descent (SGD), which is renowned for its effectiveness on large datasets, was used in the first application. This optimizer can navigate complex situations effectively as it uses randomly selected mini-batches to adjust its parameters. The next in line was Adagrad, a version that modifies each parameter's training rate according to its historical gradient. This capability is beneficial in situations where a few settings might need to be slightly adjusted. Adadelta, another adaptive optimizer, dynamically adjusts learning rates and works well in situations when a constant learning rate is not acceptable.

Finally, Adam was employed, a well-liked option because of its flexible learning rate and momentum. To offer robust and efficient parameter updates, it blends the advantages of momentum-based techniques with Adagrad. A systematic comparison of different optimizers provided insights into which approach best optimized the FNN's performance in predicting growth parameters for the aquaponics system, as shown in Fig. (**2**).

RESULTS AND DISCUSSION

The experiment conducted on the Pond 1 and Pond 2 datasets yielded impressive performance metrics through the use of various optimisation techniques, including Adam, AdaDelta, Stochastic Gradient Descent (SGD), and Adagrad. Table **1** shows that for Pond 1, Adagrad and Adam, for instance, achieved an exceptional accuracy of 0.96 at 250 epochs. Table **2** shows that Pond 2, Adagrad, and Adam also attained remarkable accuracy at 150 epochs, indicating that they were successful in fine-tuning the predictive models for both datasets. The experiment shows how optimization strategies affect the predictive power of the model. In the

Pond 1 and Pond 2 datasets, Adagrad and Adam demonstrated remarkable accuracy, demonstrating their capacity to adjust parameters for precise predictions. The best training epochs were 250 and 150, indicating that more training may not produce noticeable gains and could even cause overfitting. To accurately anticipate the growth conditions of fish in aquaponics ponds, the study highlights the significance of adapting optimization methods and hyperparameters to the unique properties of the datasets, especially Adagrad and Adam. The results demonstrate the performance of learning rates and optimizers, offering insightful information for improving aquaponics systems' fish farming techniques and leading to more effective and long-lasting methods.

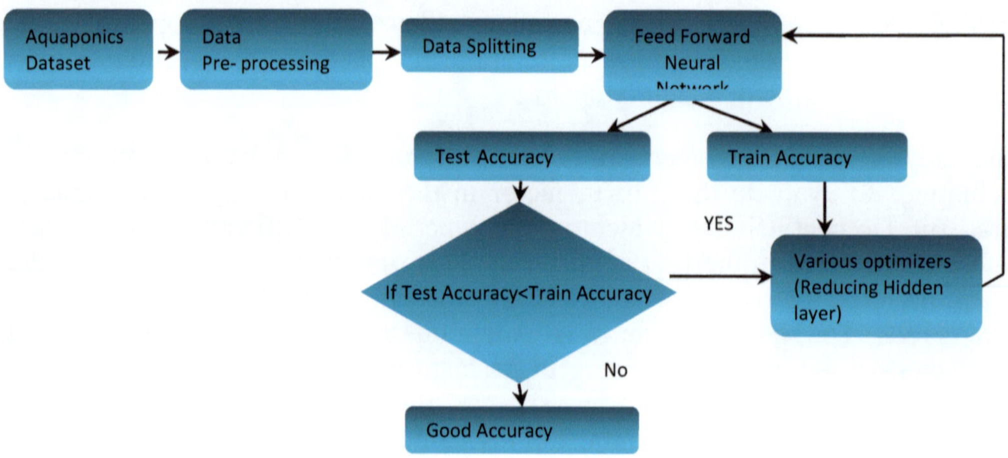

Fig. (2). Block diagram of the prediction of Fish growth factor.

Table 1. Aquaponics_pond1 dataset.

Epochs	*SGD*	*AdaDelta*	*AdaGrad*	*Adam*
0	0.10	0.12	0.16	0.18
50	0.24	0.26	0.43	0.17
100	0.43	0.36	0.51	0.68
150	0.64	0.52	0.72	0.84
200	0.72	**0.94**	0.84	0.66
250	0.52	0.66	**0.96**	**0.96**
300	**0.92**	0.72	0.74	0.18
350	0.82	0.68	0.56	0.33

Table 2. Aquaponics_pond2 dataset.

Epochs	*SGD*	*AdaDelta*	*AdaGrad*	*Adam*
0	0.10	0.14	0.12	0.11
50	0.34	**0.93**	0.36	0.39
100	0.25	0.63	0.67	0.25
150	0.54	0.45	**0.96**	**0.96**
200	0.46	0.27	0.71	0.66
250	**0.82**	0.72	0.65	0.76
300	0.72	0.82	0.42	0.18
350	0.78	0.48	0.26	0.35

The line plots in Figs. (**3** and **4**) illustrate the evolution of loss values for several optimization algorithms throughout different epochs, providing a dynamic portrayal of their performance patterns. It provides an understanding of the patterns of convergence or divergence for SGD, AdaDelta, AdaGrad, and Adam's loss evolution during training.

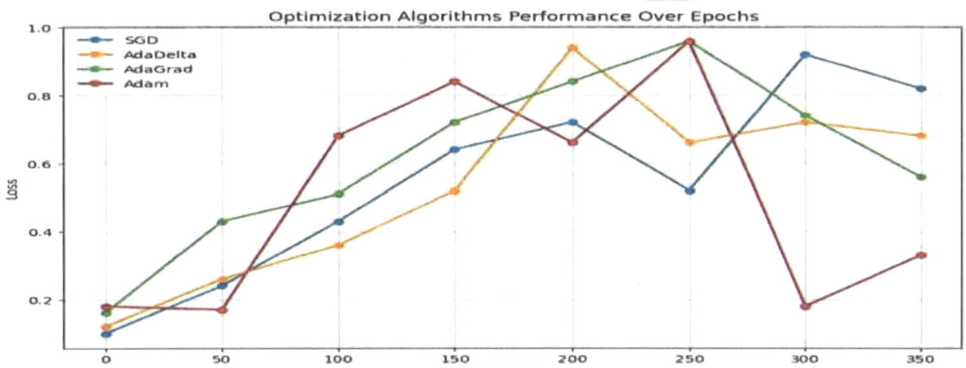

Fig. (3). Performance of optimization algorithms over Epochs in Pond1.

The matplotlib plot displays the results of four optimisation techniques over various epochs: SGD, AdaDelta, AdaGrad, and Adam. Every line reveals the distinct convergence paths of the algorithms, each of which is related to a specific one. This highlights the unique characteristics of different techniques and offers insights into their efficacy and stability in minimizing the loss function during machine learning model training. The performance of four optimisation algorithms—SGD, AdaDelta, AdaGrad, and Adam—across various learning rates and epochs is displayed in the matplotlib plot in Fig. (**5**). The lines representing individual optimizers show how convergence behaviour is affected over time by the learning rate. The graphic offers a chronological viewpoint and provides information about how well each optimizer minimizes the loss function. To optimise the training of machine learning models, this exploration is essential.

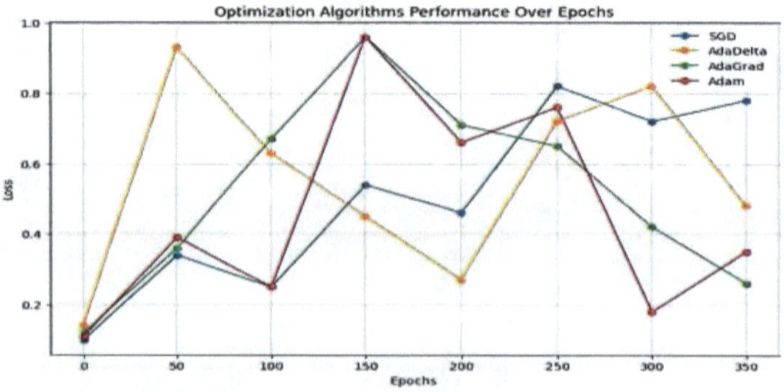

Fig. (4). Performance of optimization algorithms over Epochs in Pond2.

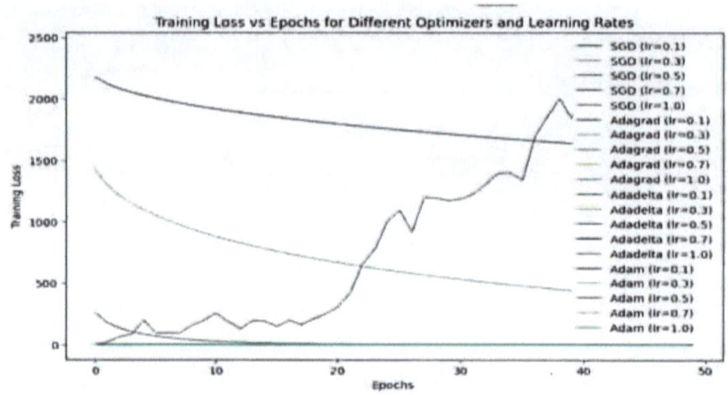

Fig. (5). Performance of optimizers with different learning rates.

CONCLUSION AND FUTURE WORK

The research work investigated the efficacy of several optimisation techniques in training a Feedforward Neural Network (FNN) to predict growth parameters in an aquaponics pond fish system. Adam was discovered to be the most effective, with greater convergence speed and accuracy. SGD and Adagrad produced good results, although with a significantly slower rate of convergence. Adadelta performed poorly in this domain. The study advances our understanding of neural networks and optimisation techniques for aquaponics.

Future research and development are suggested by the study of Feedforward Neural Network (FNN) optimizers for predicting growth parameters in aquaponics pond fish. Extending the dataset to include real-time sensor data and experimenting with advanced neural network topologies could all help increase prediction skills [25]. The combination of FNN and modern control systems may result in more efficient and sustainable operations. On-site experimental validations in real-world scenarios are crucial for the practical implementation and validation of model efficacy.

REFERENCES

[1] S. Goddek, A. Joyce, B. Kotzen, G.M. Burnell, Ed., *Aquaponics Food Production Systems: Combined Aquaculture and Hydroponic Production Technologies for the Future.* Springer International Publishing: Cham, 2019.
[http://dx.doi.org/10.1007/978-3-030-15943-6]

[2] B. Hepher, *Nutrition of Pond Fishes.* 1st ed. Cambridge University Press, 1988.
[http://dx.doi.org/10.1017/CBO9780511735455]

[3] M.A. Sheridan, "Hormonal control of reproduction and growth | endocrinology of fish growth", In: *Encyclopedia of Fish Physiology.* Elsevier, 2011, pp. 1483-1489.
[http://dx.doi.org/10.1016/B978-0-12-374553-8.00161-1]

[4] U. Knaus, and H.W. Palm, "Effects of the fish species choice on vegetables in aquaponics under spring-summer conditions in northern Germany (Mecklenburg Western Pomerania)", *Aquaculture,* vol. 473, pp. 62-73, 2017.
[http://dx.doi.org/10.1016/j.aquaculture.2017.01.020]

[5] I.H. Sarker, "Deep learning: A comprehensive overview on techniques, taxonomy, applications and research directions", *SN Comput. Sci.,* vol. 2, no. 6, p. 420, 2021.
[http://dx.doi.org/10.1007/s42979-021-00815-1] [PMID: 34426802]

[6] S. Naseri, "Effects of parameter change on neural network and deep neural network", *Dissertation,* 2021.
[http://dx.doi.org/10.32920/ryerson.14650011]

[7] Z. Zhang, "Improved adam optimizer for deep neural networks," in *2018 IEEE/ACM 26th International Symposium on Quality of Service (IWQoS)*, Banff, AB, Canada: IEEE, Jun. 2018, pp. 1–2.
[http://dx.doi.org/10.1109/IWQoS.2018.8624183]

[8] S. Keydana, *Deep Learning and Scientific Computing with R torch.* 1st ed. Chapman and Hall/CRC: Boca Raton, 2023.
[http://dx.doi.org/10.1201/9781003275923]

[9] R. Kabra, "Deep learning algorithm recommender", In: *Master of Science.* San Jose State University: San Jose, CA, USA, 2018.
[http://dx.doi.org/10.31979/etd.8gnf-yu64]

[10] N. Fatima, "Enhancing performance of a deep neural network: A comparative analysis of optimization algorithms", *ADCAIJ: Advances in Distributed Computing and Artificial Intelligence Journal,* vol. 9, no. 2, pp. 79-90, 2020.
[http://dx.doi.org/10.14201/ADCAIJ2020927990]

[11] A. Shaf, T. Ali, W. Farooq, S. Javaid, U. Draz, and S. Yasin, "Two classes classification using different optimizers in convolutional neural network," in *2018 IEEE 21st International Multi-Topic Conference (INMIC)*, Karachi, Pakistan: IEEE, Nov. 2018, pp. 1–6.
[http://dx.doi.org/10.1109/INMIC.2018.8595534]

[12] D. Choi, C. J. Shallue, Z. Nado, J. Lee, C. J. Maddison, and G. E. Dahl, "On empirical comparisons of optimizers for deep learning," 2019.
[http://dx.doi.org/10.48550/ARXIV.1910.05446]

[13] S. Postalcıoğlu, "Performance analysis of different optimizers for deep learning-based image

recognition", *Int. J. Pattern Recognit. Artif. Intell.,* vol. 34, no. 2, p. 2051003, 2020. [http://dx.doi.org/10.1142/S0218001420510039]

[14] S. Siva Shankar, B.T. Hung, P. Chakrabarti, T. Chakrabarti, and G. Parasa, "A novel optimization based deep learning with artificial intelligence approach to detect intrusion attack in network system", *Educ. Inf. Technol.,* vol. 29, no. 4, pp. 3859-3883, 2024. [http://dx.doi.org/10.1007/s10639-023-11885-4]

[15] D. Grimm, D. Tollner, D. Kraus, Á. Török, E. Sax, and Z. Szalay, "A numerical verification method for multi-class feed-forward neural networks", *Expert Syst. Appl.,* vol. 247, p. 123345, 2024. [http://dx.doi.org/10.1016/j.eswa.2024.123345]

[16] S. Banert, J. Rudzusika, O. Öktem, and J. Adler, "Accelerated forward-backward optimization using deep learning", *SIAM J. Optim.,* vol. 34, no. 2, pp. 1236-1263, 2024. [http://dx.doi.org/10.1137/22M1532548]

[17] S. Hanifi, A. Cammarono, and H. Zare-Behtash, "Advanced hyperparameter optimization of deep learning models for wind power prediction", *Renew. Energy,* vol. 221, p. 119700, 2024. [http://dx.doi.org/10.1016/j.renene.2023.119700]

[18] K. Wong, R. Dornberger, and T. Hanne, "An analysis of weight initialization methods in connection with different activation functions for feedforward neural networks", *Evol. Intell.,* vol. 17, no. 3, pp. 2081-2089, 2024. [http://dx.doi.org/10.1007/s12065-022-00795-y]

[19] N.M.H. Alamri, M. Packianather, and S. Bigot, "Deep learning: Parameter optimization using proposed novel hybrid bees bayesian convolutional neural network", *Appl. Artif. Intell.,* vol. 36, no. 1, p. 2031815, 2022. [http://dx.doi.org/10.1080/08839514.2022.2031815]

[20] A.K. Dubey, A.K. Sinhal, and R. Sharma, "Heart disease classification through crow intelligence optimization-based deep learning approach", *Int. J. Inf. Technol.,* vol. 16, no. 3, pp. 1815-1830, 2024. [http://dx.doi.org/10.1007/s41870-023-01445-x]

[21] S. Roy, R. Mehera, R.K. Pal, and S.K. Bandyopadhyay, "Hyperparameter optimization for deep neural network models: a comprehensive study on methods and techniques", *Innov. Syst. Softw. Eng.,* no. Oct, 2023. [http://dx.doi.org/10.1007/s11334-023-00540-3]

[22] J. Lim and K. Sung, "FNN (Feedforward Neural Network) Training Method Based on Robust Recursive Least Square Method," in *Advances in Neural Networks* – ISNN 2007, vol. 4492, D. Liu, S. Fei, Z. Hou, H. Zhang, and C. Sun, Eds., in *Lecture Notes in Computer Science*, vol. 4492. , Berlin, Heidelberg: Springer Berlin Heidelberg, 2007, pp. 398–405. [http://dx.doi.org/10.1007/978-3-540-72393-6_48]

[23] M. Alselek, J.M. Alcaraz-Calero, J. Segura-Garcia, and Q. Wang, "Water IoT Monitoring System for Aquaponics Health and Fishery Applications", *Sensors (Basel),* vol. 22, no. 19, p. 7679, 2022. [http://dx.doi.org/10.3390/s22197679] [PMID: 36236778]

[24] K. Arunachalam, S. Thangamuthu, V. Shanmugam, M. Raju, and K. Premraj, "Deep learning and optimisation for quality of service modelling," *Journal of King Saud University - Computer and Information Sciences*, vol. 34, no. 8, pp. 5998–6007, Sep. 2022. [http://dx.doi.org/10.1016/j.jksuci.2022.01.016]

[25] K.T. Chui, B.B. Gupta, and P. Vasant, "A genetic algorithm optimized RNN-LSTM model for remaining useful life prediction of turbofan engine", *Electronics (Basel),* vol. 10, no. 3, p. 285, 2021. [http://dx.doi.org/10.3390/electronics10030285]

CHAPTER 6

Predicting Carbon Dioxide Emissions from Green Waste Composting and Identifying Key Factors Using Machine Learning Algorithms

M. Vaishali[1,*], **M. Monika**[2], **M. Atish**[3] and **S. Rajarshi**[4]

[1] *Department of Computer Science and Engineering, Maharishi Markandeshwar Deemed to be University, Ambala, India*

[2] *Department of Information Technology, Dwarkadas J. Sanghvi College of Engineering, Navi Mumbai, India*

[3] *Department of Computer Science and Engineering, BMS Institute of Technology and Management, Bangalore, India*

[4] *School of Advanced Sciences, VIT-AP University, Andhra Pradesh, India*

Abstract: Today, one of the main concerns that the world faces is climate destruction. The hazardous gases emitted by material waste are a critical component among many that contribute to the constantly changing atmospheric conditions. Waste from agriculture is also included in this group. Controlling the production of these gases largely depends on accurately predicting the amount to be produced. The current study offers new classifiers and prediction models based on machine learning to forecast the generation of hazardous gases from agricultural waste. The Random Forest method achieved 100% classification accuracy after feature optimization and selection were applied to the gathered dataset. Logistic regression, AdaBoost, and Decision Tree ranked second, third, and fourth, respectively, with classification accuracies of 97.19%, 97.18%, and 97.14%. SVM provided the highest prediction accuracy, 98%, in regression models.

Keywords: Artificial intelligence, Environment, Green energy, Machine Learning, Prediction, Renewable energy.

INTRODUCTION

In recent decades, nations worldwide have faced significant challenges in addressing climate change, a formidable threat leading to global warming and other widespread environmental impacts. As a result, numerous academics, ecological scientists, and governmental organizations have been recommending

* **Corresponding author M. Vaishali:** Department of Computer Science and Engineering, Maharishi Markandeshwar Deemed to be University, Ambala, India; E-mail: drvaishaliwadhwa@gmail.com

Suneeta Satpathy, Sachi Nandan Mohanty & Subhendu Kumar Pani (Eds.)

various courses of action to overcome these difficulties. For example, the government is actively attempting to raise public awareness of the seriousness of environmental issues and defacement [1]. In addition, scientists and researchers are working to implement solutions that will reduce the production of harmful gases. To mitigate environmental harm, efforts are also being made to convert these hazardous gases into beneficial energy. Along similar lines, some scientists have suggested transforming greenhouse gases into energy that may be used, such as electricity [2]. Approximately three-fourths of all greenhouse gas emissions originate from the urban sector, which is the primary source of these emissions. Furthermore, this percentage may increase due to the growing urban population, necessitating effective corrective action. While there is no way to prevent or manage agricultural waste, steps may be taken to maximize its use, according to some experts and environmentalists working in this area [3, 4]. For example, agricultural waste can be converted into biogas, which can then be used to generate electricity. Typically, agrarian waste releases CO_2, CH_4, N_2O, and SO_2 gases. These gases are then converted by microorganisms into biogas within biogas plants *via* chemical reactions. The resulting biogas is a renewable soure of energy that can help produce electricity.

Furthermore, it has been noted that, of all the gases, the emission of CO_2 is the highest; this gas is a key contributor to environmental harm. Therefore, it is essential to forecast CO_2 emissions from agri-waste since they significantly impact the prediction of biogas and environmental health [5]. It is important to note that recycling the material helps protect the environment and improves the economy. Therefore, considering the various advantages of generating biogas from agricultural waste, it is essential to capitalize on the technological advancements in this domain. The authors of this work have made an effort to utilize Machine Learning (ML) and Artificial Intelligence (AI) models in this field, building on previous work in this direction [6]. ML models are a practical option in this field because of their enormous potential, which has been demonstrated over many years in information forecasting and categorization. The current efforts aim to evaluate the predictive accuracy of various machine learning models for CO_2 emissions and biogas production. The authors have employed diverse categorization and regression methods on datasets encompassing 192 nations over ten years. Several ML algorithms, such as Random Forest, KNN, Naïve Bayes, Decision Tree, Logistic Regression, SVM, Linear Regression, and ADA Boost, have been used in the experiments. A comparison analysis is then conducted to identify the most effective ML model for this particular case. This research aims to estimate the predictor class using the most widely used regression models and machine learning classifiers, after gathering a dataset with sufficient labeled variables. The study is motivated by the fact that most existing methods, which have been utilized in other studies, have failed to achieve high accuracy levels for the same class of situations [1, 5]. Furthermore, existing investigations do not utilize EDA to gain insight into the problem's parameters and their impact on the obtained outcomes.

The current work is divided into many components. In this case, section 2 outlines the previous relevant research projects completed by various researchers. The implementation process is expounded upon in Section 3. Results are covered in Section 4, and Section 5 concludes with a presentation of future work.

LITERATURE SURVEY

A study suggests that CO_2 is a crucial component produced during the composting of green waste. Its authors researched how to predict CO2 during this process. A comparative analysis of six different methods was conducted. Before applying machine learning (ML) models, preprocessing was performed, which included the removal of outliers. The results of the comparative study show that the random forest algorithm outperforms other models in terms of accuracy, achieving an accuracy of 88% [1].

Additionally, the authors of another study propose using renewable energy sources since they drastically reduce waste and safeguard the environment [2]. For prediction purposes, their study proposes a hybrid model that combines Random Forest with Long Short-Term Memory (LSTM) [10]. The accuracy of this hybrid model is 20% higher than its conventional analytical models. Another study also establishes the use of ML models in the present situation. The authors conducted research with two primary aims: first, to develop a reliable method for estimating waste production; and second, to compare various quantification techniques. To achieve these objectives, they employed four specific models: artificial neural networks, decision trees, grey models, and multiple linear regression [4]. The authors of another study employed a deep neural network to estimate the quantities of SOx and NOx released during the conversion of coal to energy, following a similar approach [5].

Numerous researchers have addressed this problem and proposed several solutions in light of the pervasive and profound effects that toxic gases have on the environment [7 - 9]. A study has proven that an effective model for decision-making is necessary to improve biogas production [3]. Given the input waste values, a graphical user interface is needed to forecast the amount of biogas produced. To achieve this, the authors used various machine learning models, including logistic regression, random forest, support vector machines, gradient boosting, and kNN regression. With an accuracy of 87% for test data, kNN was found to have the best accuracy for evaluating the model efficiency at the Hainan biogas facility.

PROPOSED METHODOLOGY

The primary objective of this work is to predict and analyze hazardous gases generated from green waste. Additionally, maximizing accuracy and decreasing computing costs have been prioritized. Machine learning algorithms have also employed various optimized data analysis techniques, including feature selection, exploratory data analysis, and data pretreatment.

After data collection, seven Machine Learning classifier models were implemented to identify the most efficient predictive models by comparing their efficiency using measures such as F1 score, Accuracy, Precision, and Recall. Finally, the efficiency of the proposed WQI model is evaluated using the ROC technique, a validation curve, and a confusion matrix for different predictive models.

Dataset Collection

The dataset utilized in this research was gathered from (https://www.kaggle.com/datasets/die9origephit/greenhouse-gases-and-air-pollutant-emissions). A dataset of 192 countries was compiled, comprising country-specific data on greenhouse gases from the past decade. The dataset contains only continuous variables, as illustrated in Table **1**.

Table 1. Agri-waste emission parameters.

Feature Name	Description	Feature Type	Per Capita Range
CH_4	Methane	Numeric	0.001 kg – 14.28 kg
CO_2	Carbon Dioxide	Numeric	0.02 kg – 47.28 kg
N_2O	Nitric Oxide	Numeric	0 kg – 11.28 kg
NOx	Nitrogen Oxide	Numeric	0 kg – 144.19 kg
SO_2	Sulfur Dioxide	Numeric	0 kg – 247.93 kg
CL	Cellulose Content	Numeric	0.16 kg – 61.57 kg
Hemi CL	Hemicellulose	Numeric	0.18 kg – 57 kg

Data Preprocessing and Filtering

The dataset consists only of continuous values, as illustrated in Table **1**. Nevertheless, the dataset includes duplicate and missing values, which are a primary cause of performance reduction. For continuous variables, the mean is used to replace missing values; for discrete variables, the median is used as an alternative. The data is substituted with numbers (*e.g.*, NA is replaced with 0, OK is replaced with 1, Low is replaced with 0, *etc.*) for accurate analysis. Furthermore, a few irrelevant elements in the dataset also impact accuracy because they don't improve the prediction. As will be covered in the following section, the suggested method addresses these qualities by employing feature extraction and feature selection [1].

Feature Extraction

A "feature extraction" or "feature elimination" method lowers the model's dimensions while retaining the most crucial features for additional processing. This procedure minimizes information loss while reducing the quantity of redundant information in data analysis and the number of features in a dataset [2]. As a result, it enhances model efficiency and lessens computing effort. We have employed Principal Component Analysis (PCA) as a feature extraction tool in our methodology. PCA uses the least correlated features for analysis to determine the correlation between features. This procedure reduces high variance, which in turn reduces overfitting [3].

Feature Selection

Our model performs feature selection after feature extraction to achieve more precise results. The most reasonable and promising features are selected from the reduced feature set, while irrelevant features are ignored, according to the feature selection process. Lasso regularization is one of the most compelling feature selection methods [4]. The Lasso regularization method operates based on the total sum of its coefficients. This technique prevents over-fitting by adding a penalty to various model parameters. Hence, the coefficients of the grassy characteristics eventually approach zero, eliminating those features from the model and leaving just the most crucial ones. The features are pre-filtered by the Gini Index (GI) and Information Gain (IG) parameters during the Decision Tree modeling procedure [5].

Applying Machine Learning Classifiers for CO_2 Prediction from Agri-waste

As mentioned below, many machine learning classifiers and regression models have been applied and examined for the current research [6].

Linear Regression

Linear regression is a method in supervised machine learning that predicts the linear relationship between independent variables and a dependent variable. Stated differently, the linear combination of the input and independent variables can be used to predict the target variable.

Logistic regression is one of the best supervised learning techniques for binary classification. Furthermore, when data points cannot be separated linearly, they are employed in non-linear classification. It forecasts the likelihood of the output class using a sigmoid function. The output gradually changes until the stopping criteria are satisfied since the sigmoid function's value varies slowly [6].

Naïve Bayes

The Naïve Bayes classifier is a supervised learning method for categorizing high-dimensional text datasets. It uses the following technique to predict or categorize the target variable based on the likelihood of individual features:

$$Y = armax\ P(y) \prod_{i=1}^{N} P(x_i|Y),$$

P(y) is the posterior probability, and (xi,y) is the conditional probability.

Decision Tree

A categorization technique based on a top-down tree, known as a decision tree, determines the optimal classifier by calculating entropy and information gain. DT is a powerful method for classifying data into multiple classes. Using our dataset, we applied the DT model to predict classifications of Normal, Mild, Moderate, Severe, and End-stage.

KNN

The training set's closest data points are classified by the KNN classifiers [7]. Our model uses an Euclidean distance-based approach to determine the nearest data points. The KNN algorithm's output is highly dependent on the value of K. While a smaller value of K may increase output noise, a larger value of K may decrease accuracy.

SVM

A popular supervised learning technique for classification and regression issues is SVM. SVM is renowned for being a "model-free" approach because it is independent of the distribution of the parameters. Every data point is shown as a vector with n dimensions. The SVM model divides them into two discriminant classes using an n-1-dimensional separating hyperplane. The goal is to maximize the separation between the data points and the hyperplane. Studies show that SVM is as accurate as, or more accurate than, logistic regression in classifying diseases.

Ensemble Modeling

Ensemble modeling is a machine learning technique that leverages the capabilities of numerous machine learning models to maximize accuracy [8, 9]. The two popular ensemble modeling methods employed in this study are Random Forest and AdaBoost.

Random Forest

Among various classifiers, Random Forest is an ensemble learning method that demonstrates the highest accuracy. It employs the bagging technique, which is considered a replacement for random sampling of data. Several decision trees with randomly chosen feature subsets are fed these random samples. Ultimately, the algorithm makes its final prediction through a voting process. RF employs the Gini index and information gain to pick the feature subset optimally.

AdaBoost. AdaBoost is an additional ensemble learning method that utilizes input parameter weights. AdaBoost employs a boosting technique to assign weights to specific parameters, which are subsequently selected based on these weights. In contrast, RF uses bagging to draw random samples from the dataset. While several classifiers (often decision trees) operate simultaneously in reinforcement learning, distinct classifiers operate sequentially in learning.

Identifying Key Factors using Exploratory Data Analysis (EDA)

EDA contributes to a deeper comprehension of the problem and the data factors. EDA is an effective technique for drawing conclusions and eliminating bias. Therefore, exploratory data analysis is a logical next step. In this study, a univariate analysis is conducted. The dataset has been separated into numerical and categorical variables for easier comprehension. An attempt has been made to use the KDE plot and box, as seen in Fig. (**1**), to investigate the distribution of our variables. The KDE plot facilitates the visualization of data concentration, distribution shape, and variable distribution. However, a Boxplot can show outliers, Quantiles, and the median.

Data visualization shows that the two leading gases emitted during chemical reactions between agricultural waste and the atmosphere are CO_2 and CH_4. These, therefore, represent the primary markers for the classification issue. Since CO_2 is the primary cause of climate damage, we have used it as the term for this effort. The graphs above show a few outliers, most of which fall on the lower end of the edge spectrum and can, therefore, be disregarded. The remaining gases are concentrated close to zero, indicating that very little is created during the emission process.

RESULTS AND DISCUSSION

This section explains the outcomes of the simulation procedure. The dataset's source is Kaggle. The dataset includes plant waste gas emissions from 190 countries over the past decade. During data preprocessing, approximately 100 values were identified as inconsistent; these were either removed or corrected. The next stage is feature optimization after data filtering. During the emission process, agri-waste releases about six gases: CO, CO_2, CH_4, NO_2, SO_2, and NOx. As was covered in Section 1, CO_2 is the single gas that contributes most to the production of green energy. Nevertheless, the simulation is run for the entire dataset to compare and then for a subset of features.

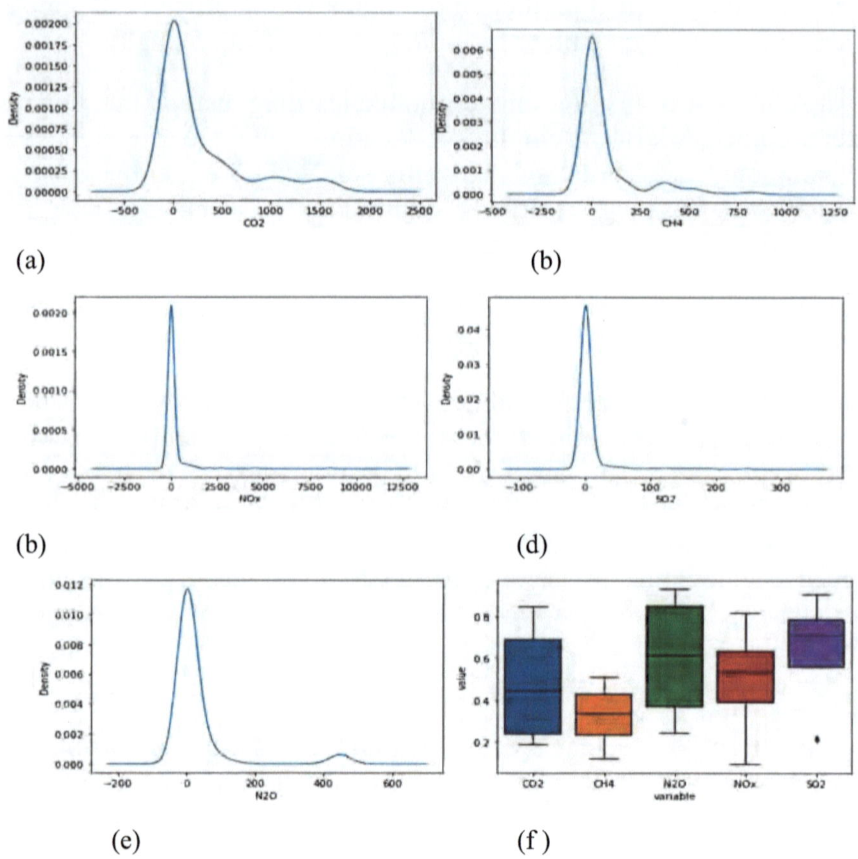

Fig. (1a-f). KDE plots of all 11 optimized variables.

This study on CO_2 prediction, which may be further utilized for generating greenhouse gases, is motivated by the collected dataset, which indicates that the emission of CO_2 from plant waste is the largest among other greenhouse gases. Nonetheless, this research holds great potential for forecasting global warming.

Fig. (**2**) shows the release of CO_2 in the top 10 countries around the globe in the year 2020.

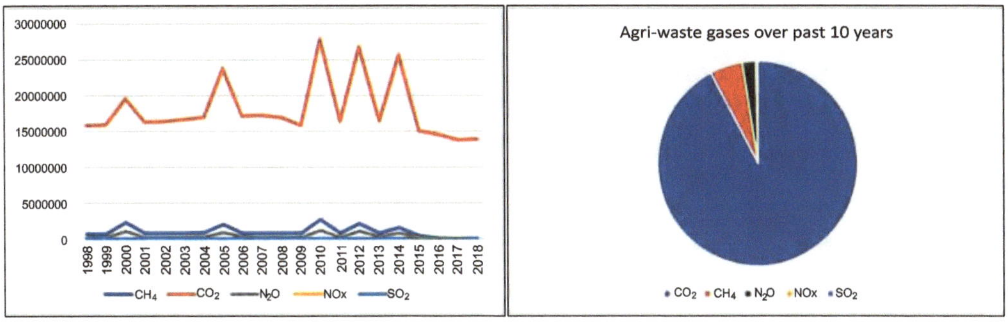

Fig. (2). Emission of Greenhouse gases in the past 10 years.

Here, Fig. (**2**) also clarifies that the most common gas generated during green waste is CO_2. Fig. (**3**) also shows the generation of CO_2 by country. Sensitivity (1), Specificity (2), F1 Score (3), Accuracy (4), and AUC are some of the evaluation metrics used to compare the efficiency of the various machine learning models employed by the authors in this work. These measures have the following mathematical formulations [11, 12].

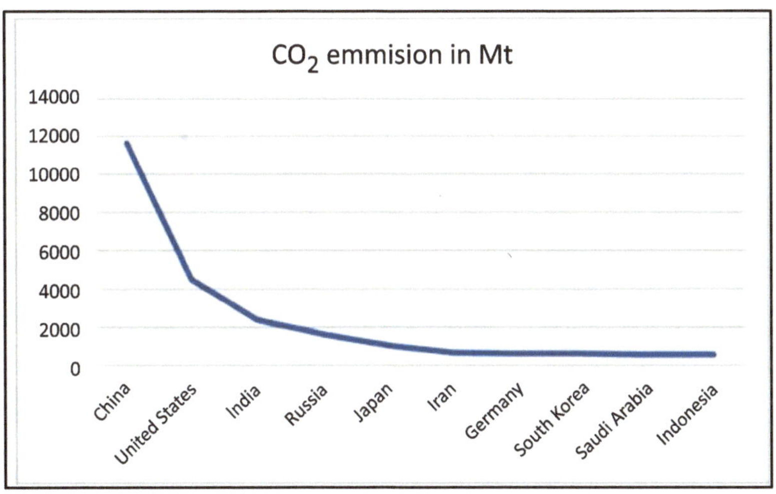

Fig. (3). CO_2 emissions of the 15 Largest countries in 2020.

$$Sensitivity = \frac{TP}{TP+FN} \times 100 \quad (1)$$

$$Specificity = \frac{TP}{TP+FP} \times 100 \quad (2)$$

$$F1\ Score = 2 \times \frac{Sensitivity \times Precision}{Sensitivity + Precision} \times 100 \quad (3)$$

$$Accuracy = \frac{TP+TN}{TP+FP+TN+FN} \times 100 \quad (4)$$

In all the above equations, TP, FP, TN, FN, represent True Positive, False Positive, True Negative, and False Negative, respectively. Finally, *N* indicates the number of values in the dataset. The gathered dataset is trained and tested using ML algorithms to validate machine-learning classification techniques for biogas production. The accuracy and efficiency of these algorithms were assessed and compared. With a 100% accuracy rate on the validation set, Random Forest is the most accurate classifier, followed by Logistic Regression with a 97.91% accuracy rate. AdaBoost and Decision Tree demonstrated strong performance, achieving accuracy rates of 97.89% and 97.84%, respectively. Nonetheless, Naïve Bayes has an accuracy of 89.4%. Moreover, it can be seen in Fig. (**1**) that after removing the outliers, the variance is reduced slightly compared to both datasets, *i.e.*, before and after feature extraction. The ensemble modeling-based Random Forest algorithm outperformed after outliers were removed.

The simulation results for the different performance metrics are shown in Table **2**. In terms of accuracy and estimated error, Naïve Bayes, KNN, Random Forest, and Adaboost have demonstrated the highest levels of efficiency. But when it comes to binary classification, KNN, RF, and Adaboost take longer than Naïve Bayes. In the event of a larger dataset, this might rise. Consequently, Naïve Bayes provides the best results for binary classification.

Table 2. Results of different classifiers obtained post feature reduction.

Classifier	Precision	Recall	F1 Score	RMSE(%)	Accuracy
Naïve Bayes	0.88	0.91	0.90	0.12	89.2
KNN	0.95	1.0	0.97	0.07	96.5
Log. Reg.	0.97	1.0	0.98	0.0087	99.1
Decision Tree	0.96	1.0	0.98	0.035	97.89
Random Forest	1.0	1.0	1.0	0	100
AdaBoost	0.96	1.0	0.98	0.06	97.84
SVM	1.0	0.96	0.98	0.04	98
SVM(Poly)	0.96	1.0	0.90	0.08	92

(Table 2) cont.....

Classifier	Precision	Recall	F1 Score	RMSE(%)	Accuracy
SVM(rbf)	0.90	0.83	0.89	0.24	87
Linear Reg	0.94	0.80	0.89	0.14	91.3

Table **2** shows that Naïve Bayes, KNN, Random Forest, and Adaboost have demonstrated the highest efficiency in terms of accuracy and error estimation; however, KNN, RF, and Adaboost take longer than Naïve Bayes for binary classification, which may increase with a larger dataset. As a result, Naïve Bayes provides the best results for binary classification. RMSE determines the classifier's efficiency by measuring the gap between the actual and predicted values of the CO_2 prediction.

CONCLUSION AND FUTURE SCOPE

Since the process of producing biogas from agricultural waste is still in its infancy, protocols have not yet been established. As a result, effective prediction models are necessary to forecast the volume of gases produced throughout the biogas generation process. Using machine learning models, decision-makers can better understand the various variables and avoid releasing hazardous chemicals into the atmosphere. The authors of this publication have proposed a machine learning-based prediction model to help achieve this goal. It has been observed that the proposed model is effective in predicting the release of hazardous gases from agricultural waste. The proposed model can be helpful to policymakers as they strive to protect the environment from harmful emissions during the biogas generation process. The suggested work might be expanded to cover several approaches to data reduction methods, such as Principal Component Analysis (PCA). Practical techniques for reducing data, such as entropy and the Gini Index, will also help enhance the model's performance.

REFERENCES

[1] R. Kaladevi, S. Saidineesha, P. Keerthi Priya, K.M. Nithiya, and S. Sai Gayatri, "Chatbot for healthcare using machine learning", *2023 International Conference on Computer Communication and Informatics (ICCCI),* pp. 1-4, 2023.Coimbatore, India. [http://dx.doi.org/10.1109/ICCCI56745.2023.10128261]

[2] L. Medeiros, T. Bosse, and C. Gerritsen, "Can a chatbot comfort humans? Studying the impact of a supportive chatbot on users' self-perceived stress", *IEEE Trans. Hum. Mach. Syst.,* vol. 52, no. 3, pp. 343-353, 2022. [http://dx.doi.org/10.1109/THMS.2021.3113643]

[3] G.A. Santos, G.G. de Andrade, G.R.S. Silva, F.C.M. Duarte, J.P.J.D. Costa, and R.T. de Sousa, "A conversation-driven approach for chatbot management", *IEEE Access,* vol. 10, pp. 8474-8486, 2022. [http://dx.doi.org/10.1109/ACCESS.2022.3143323]

[4] S.S. Chua, Y.M. Austria, M.G. Intal, and J.J.B. Ong, "SERENE: An online support system for depression", [http://dx.doi.org/10.1109/ICBIR52339.2021.9465876]

[5] A. Gupta, and N. Basit, "Ranking actionable mental health advice through a personalized video recommendation system", [http://dx.doi.org/10.1109/ICICoS56336.2022.9930526]

[6] A. Teles, "Mobile mental Health: A Review of Applications for Depression Assistance", [http://dx.doi.org/10.1109/CBMS.2019.00143]

[7] V. Laijawala, A. Aachaliya, H. Jatta, and V. Pinjarkar, "Classification Algorithms based Mental Health Prediction using Data Mining", [http://dx.doi.org/10.1109/ICCES48766.2020.9137856]

[8] S. Abu Noman Siddik, B.M. Arifuzzaman, and A. Kalam, "Psyche Conversa - A Deep Learning Based Chatbot Framework to Detect Mental Health State", [http://dx.doi.org/10.1109/ICoICT55009.2022.9914844]

[9] R. Raut, S. Kolambe, P. Borkar, A. Shingade, and M. Gulhane, "Depression Therapy Chat-Bot using Natural Language Processing", *Int J Intell Syst Appl Eng,* vol. 12, no. 2s, pp. 181-187, 2023.

[10] A. Macharla, and G. Pradeepini, "A novel framework for the diagnosis of Parkinson's disease using transfer learning with RESNET50 and SVM classifier", *Indones. J. Electr. Eng. Comput. Sci.,* vol. 32, no. 2, pp. 877-866, 2023. [http://dx.doi.org/10.11591/ijeecs.v32.i2.pp877-886]

[11] S. Rasheed, "G. Kiran Kumar, D. Malathi Rani, MVV Prasad Kantipudi, and M. Anila. "Heart Disease Prediction Using GridSearchCV and Random Forest."", *EAI Endorsed Trans. Pervasive Health Technol.,* p. 10, 2024.

[12] A. M, G.K. Kumar, D.M. Rani, M.V.V. Prasad Kantipudi, and D. Jayaram, "An LSTM based DNN Model for Neurological Disease Prediction Using Voice Characteristics", *EAI Endorsed Trans. Pervasive Health Technol.,* vol. 10, 2024. [http://dx.doi.org/10.4108/eetpht.10.5424]

CHAPTER 7

Can Ethiopian Cooperative Coffee Farmers Predict Financial Management Decisions Through Human Capital and AI Mediation?

Mershaye Birhane[1], **Shashi kant**[1,*], **Brehanu Borji**[2] and **Chalchisa Amentie**[3]

[1] *Department of Management, Bule Hora University, Bule Hora, Ethiopia*

[2] *Department of Management, Hawasa University, Hawasa, Ethiopia*

[3] *Department of Management, Ethiopian Civil Service University, Addis Ababa, Ethiopia*

Abstract: The central theme of the chapter was to demonstrate how Ethiopian cooperative coffee farmers can make informed financial management decisions through the mediation of human capital and AI. The investigators employed a cross-sectional examination using the multistage method with the test group. An Exploratory Factor Analysis (EFA) was conducted to verify whether the data were sufficient. Eigenvalues were utilized to explore the principal proxies in the dataset. A Confirmatory factor analysis was used to assess whether the manifested determinants could appropriately describe the validity of the construct. SEM was used to conduct mediation analysis and assess model fitness. According to the results, the investigation revealed a substantial association, as the p-value was below the algebraic threshold for data adequacy, and the sphericity was 0.893. The chi-square was below 3.0, which the investigators found indicative of model adoption. By Tucker Lewis's fitness, the model was manifested as a fit. The predictive ability of farmers has increased by 75% when coffee cooperatives practice human capital mediation. Consequently, there is a fractional conflict between the sustainability of cooperatives and their management of financial methods, brought about by human capital and AI applications.

Keywords: AI, Coffee cooperatives, Financial management Strategies, Firm performance, Human capital.

INTRODUCTION

The organization's success is influenced by the growth of human capital and technological foundations, such as Artificial Intelligence (AI) [1]. Therefore, human capital development requires careful consideration of financing sources, technological advancements such as AI, and informed investment choices [2]. The

* **Corresponding author Shashi Kant:** Department of Management, Bule Hora University, Bule Hora, Ethiopia; E-mail: skant317@gmail.com

Suneeta Satpathy, Sachi Nandan Mohanty & Subhendu Kumar Pani (Eds.)

company will profit by improving its human capital and artificial intelligence (AI)-based management system and selecting the relevant asset classes to invest in and financing sources [3]. The corporation is expected to see an increase in profits over time from its investments in AI and human resources. Businesses with good investment potential can benefit from rising stock prices and promising prospects, increasing the firm's worth.

Processing businesses enhance their financial success through modern cost management strategies that leverage AI and human capital development [4]. In cooperative settings, adopting effective asset management *via* investments in both human capital and Artificial Intelligence (AI) is expected to boost firm sustainability and market share. This is achieved through improved capital structure predictability, which enables better decision-making, facilitates scaling, and enhances customer satisfaction [5].

Internal controls of cooperatives ensure that companies manage their capital structure in compliance with rules and regulations [6]. Robust internal controls have the potential to enhance operational performance and provide precise financial reporting in the event of internal or external audits [7]. Costing, resource management, inventory agility, and investment in AI and human capital are all aspects of financial management procedures [8]. Community-based businesses are carefully designed to enhance the usefulness, value, and desirability of specific offerings through investments in both human and artificial intelligence [9].

THEORETIC ESTABLISHMENT

Workforce quality is a key component of the broader concept of AI and human capital, encompassing various other aspects [10]. Because they are seen as investors in the company, staff are rewarded with social wealth and the expected return on their assets [9, 10]. Based on a study, three main viewpoints have been presented regarding human capital [11]. According to the principal perspective, capital in the form of humans is the outcome of sharing; as such, its value is assigned to improving physical capabilities, innate intellect, and obtaining knowledge and skills [12, 13].

The Resource-Based View (RBV) theory relates to investing. The subsequent concept of partial output takes human capital into account. It is defined as the specialized AI knowledge, AI knowledge base competence, operational experience, and the expertise of managers, technicians, and inventors [14]. Notably, the perspective on overall output characterizes human capital as the culmination of an individual's physical prowess, knowledge, intellect, and other abilities used in production [15]. The last point of view, which holds that a person's productive potential is derived from their human capital, has gained significant support. It is thought that technical and managerial personnel are not the only ones with access to human capital [16].

EMPIRICAL LITERATURE REVIEW

Fiscal Supervision and Business Success

Every institute needs its management to be prepared to take risks and explore new possibilities to succeed. Providing the required human effort, which is rewarded, entails risk [15]. In contemporary financial management, theoretical and statistical techniques are applied [16]. Employees and other interested parties are invited to provide their comments on prospective subjects of interest during the early phases. The viewpoints inform certain judgments by reviewing all the replies [17].

Financial management is an approach by which an institution uses a range of logical skills to scrutinize historical, current, and future financial results to make informed decisions about its overall progress [18]. The expansion of the economy depends on the business climate [19]. It is precarious to recognize that an economy's future depends entirely on how well its institutions are managed. The financial management plan should be prioritized before deciding on the expected returns [20]. It is divided into short and strategic fiscal prospects [19]. But eventually, both of them reach a point when capital returns surpass expectations [21]. These fundamentals are combined to depict the nature of practices used to manage finances [22]. Including investment in humans, along with balanced investment decisions, leads to optimal conditions [23].

H1: Fiscal Supervision has a statistically noteworthy association with business success.

Management of Finance and Investment in Humans

Human capital refers to an employee's skill and ability value in terms of money [24]. Employers value loyalty, punctuality, skills, intellect, abilities, training, and good physical and mental health [23]. Finance distributes capital to meet corporation objectives by managing spending and revenue. HR recruits, encourages, and inspires employees to attain similar objectives. This department is typically the most expensive part of an organization's HR [25].

In more modern times, the prevailing philosophy, shaped by diverse ideologists, was considered by Becker and Schultz in the 1960s to emphasize the value of human talents [24, 25]. Like any other form of capital, Schultz claims that human capital can be employed to enhance both the quantity and quality of production. This would require investing funds in training, employee development, and enhanced benefits [23].

Although they rely on spending on training talents and knowledge through awareness, human capital investments are calculable. HR managers can calculate total profitability before and after investments [25]. To calculate any possible Return On Investment (ROI), the entire earnings are considered dividends [24].

H2: Management of finances has a statistically substantial link with human investment.

Financial Management and AI

Artificial Intelligence (AI) tools, such as predictive analytics and machine learning, can improve decision-making in financial management by automating procedures and offering insightful information [24]. Financial organizations can evaluate and manage risks more effectively thanks to Artificial Intelligence (AI), which can analyze vast volumes of historical data to identify patterns and trends. Algorithms for machine learning can aid in fraud detection, credit scoring, and identifying potential market risks [21, 23]. Algorithms driven by Artificial Intelligence (AI) may evaluate sentiment in social media, news, and market data to deliver real-time insights and suggestions for investment choices [24].

AI systems can automate trading procedures by evaluating market data, placing trades, and adjusting investment portfolios according to preset standards [22]. The primary goal of high-frequency trading, which primarily utilizes AI algorithms, is to capitalize on market inefficiencies and generate profits by executing trades quickly [23]. AI systems can find and stop fraudulent activity [25].

Over time, machine learning models can become increasingly successful as they continually learn and adapt to new fraud trends [26, 27]. By examining past data, market patterns, and economic indicators, Artificial Intelligence (AI) can help with financial planning and forecasting [28]. To guarantee impartial and equitable decision-making, AI systems must be transparent, accountable, and explainable [29].

H3: Management of Financial has a link using AI.

Human Capital and Organization Performance

Human capital comprehension is another essential component for a company's efficacy and durability [22]. Skills gained through training augment the staff's decision-making abilities, enhancing their potential for efficiency and effectiveness in developmental activities [24]. Spending money on training helps draw and retain faculty, as workers can see the opportunities for advancement and development within the company [24]. Job happiness and improved organizational and personnel performance are related [25].

It has been determined that one of the most important factors influencing an organization's performance is its human capital, comprising the productive capacities, expertise, and AI knowledge bases of its members [26]. Previous research has demonstrated that leverage may be obtained through human capital. Human capital is best understood holistically, although some studies focus only on specific facets of it [27]. However, most prevalent studies on this topic typically concentrate on large enterprises [28].

Human capital is essential for business growth; subsequently, it helps improve staff attributes and the process of knowledge acquisition [26]. The capacity to learn has become crucial for enhancing an organization's ability to acquire, disseminate, and apply knowledge, thereby facilitating and improving its business operations [27]. It also helps companies adapt to a volatile and unpredictable business climate. Enhanced mechanisms for sharing and absorbing information across organizational departments lead to better product or service development [28].

H4: Human capital has a substantial arithmetical link with Institute success.

AI and Organization Performance

Artificial Intelligence (AI) may improve many facets of operations, decision-making, and customer experiences, substantially influencing an organization's success [28]. Taiebat *et al.* (2018) posit that repetitive processes are time-consuming and take a long time to complete. Still, through AI technology, these activities can now be automated and done more efficiently [29]. Resources allocation, avoiding errors, and process rationalization advantages may be achieved if organizations use AI systems [30]. Higher organizational efficiency, shorter time-to-market, and reduced costs are potential benefits of such enhancement [31].

Organizations can use AI for data analysis to improve their competitiveness, discover new opportunities, and enhance operations [32]. Consumer experiences may be personalized with the help of AI technologies, including natural language processing and machine learning [34]. Such systems may assess client information, tastes, and actions to recommend specific products, ideas, and aids [33]. Organizations may increase consumer satisfaction, loyalty, and performance through personalization [35].

AI may be a great asset in risk management because it can review information and find patterns and potential risks. Machine learning algorithms can identify patterns associated with fraud, cyber security violations, or operational risks [36]. Organizations may reduce risks, minimize losses, and guarantee business continuity by employing AI for risk management [37]. AI may help with idea generation, market research, and product optimization, leading to innovation and product development [38].

H5: AI has a statistically significant influence on the Performance of an Organization.

Human Capital Mediation

According to the RBV approach, procedures such as teaching are resources used to improve business staff presentation, as they ensure that employees have the abilities and knowledge required to meet expected goals [33]. Positive attitudes are also fostered by human capital strategies [32]. Education and training of staff within the company are necessary to administer, control, and maintain a sufficient level of AI while maximizing profitability or performance [34].

Inclusive training programs create a win-win situation for businesses, committed staff, and labor productivity in economies [35]. Human capital accumulation yields long-term productivity returns on present investment, just as the accumulation of social capital [36]. Investment strategies serve as a connection between firm sustainability and committed staff inclusion [37]. It helps to make it possible to mobilize the expenses on humans and to allocate its resources in the most lucrative manner possible [36].

H6: Human Capital acts as a Mediator.

Mediation of AI between FM strategies and OP

The function of AI technologies in affecting the link between financial management and firm sustainability is considered the "mediating role of AI for strategies to manage assets." Potentially, AI may serve as a mediator by enhancing the effectiveness and productivity of financial management techniques used in business organizations; consequently, it may improve corporate performance [31]. This, in turn, enables AI to offer services such as automating time-consuming and repetitive FM operations, including data input, reconciliation, and report production [32].

With AI, financial management procedures can be enhanced, as they are more optimized; in this way, more finance specialists may devote their time to more critical issues [39]. As a result of cost reduction and high production rates, this one was enhanced [33]. Regression provides essential information on decision-making in the financial realm [34]. While utilizing Natural Language Processing (NLP), AI technology can identify patterns, trends, and correlations in economic data, which can be challenging for human analysts [36].

AI technologies, such as chatbots and virtual assistants, could enhance the satisfaction of new clients by providing personalized financial advice and assistance [38]. Organizations can enhance customer satisfaction and loyalty by customizing their financial products and services using AI to understand consumer preferences, behavior, and needs [39]. As a result, AI may serve as a mediator by boosting risk management and helping strategic planning [34].

H7: The mediation of AI was found between FM strategies and OP.

RESEARCH METHODOLOGY

Description of the Study Area

One of Ethiopia's most important areas, Oromia, generated 3,099,856.33 quintals of coffee in the current fiscal year. In 1999, the Federation of Oromia Coffee Producers Cooperatives was founded, representing Ethiopia's green coffee growers [37], and OCFCU experienced its most significant rise. There are 250 cooperatives in all, with 250,000 members [28].

Study Population, Sample Size, and Data Sources

Using an open-ended instrument, 425 participants in associations of coffee agriculturalists provided the first-hand information. The population under study consisted of the 33 principal farmers' unions of coffee in six districts that produce coffee within the study zones. These unions have been working here for over five years. When doing a multiple regression, utilizing an acceptable percentage of 10-by-one events was suggested. Similarly, Hair *et al.* (2010) asserted that a 10 by 1 sample size is appropriate [21]. Consequently, balanced panel time series data from 10 consecutive years of reviewed financial statements were used to obtain the secondary data.

Sampling Practices

Random sampling from extremely massive groups with a range of similar geographies is made feasible with cluster sampling [35]. Cost reduction through improved sample efficacy is one of the main objectives of cluster analysis. Furthermore, items with a big data set have a substantially lower chance of being chosen from the final sample. Specific strata were taken, and various cluster sizes can be used to lessen this issue [6].

Data Analysis Tool

With SPSS AMOS version 26, the data analysis was carried out. The complete statistical applications encompass typical multidimensional methods [30].

Data Adequacy Test

The data under Table **1** were found to be sufficient for factor analysis by the team of investigators using statistical methods, and the KMO diagnosis was 0.893. The matrix was scanned by using the data adequacy examination. Research indicates a substantial association in the data; subsequently, a KMO worth of 0.893 was found to be greater than 0.5, and the criterion was below 0.05. Determinant adequacy revealed a considerable relationship between business performance and financial management techniques, resources, AI, and human capital.

Exploratory analysis for the reduction of factors was depicted by investigators in Table **2** to exclude minor proxies inside the Cross matrix. PCA was used to reduce minor determinants. PCA Researchers discovered that the initial Eigenvalues of financing, AI, and human capital were all larger than 1. They were able to correlate the degree of variability throughout the research dataset with each of the principal components (eigenvectors) by using a statistical measure [36]. Therefore, while validating factory analysis, asset AI and AI and human capital are considered.

Table 1. KMO and Bartlett's test.

Adequacy of Sample. Measure (KMO)		0.893
Sphericity Bartlett's Test	CMIN/DF	129831.376
	DF	1431
	Sig.	.000

Confirmatory Factory Analysis

Confirmatory factor analysis (CFA) is used to verify the structure of factors like FM, artificial intelligence (AI), and human capital. Statistical demonstration of a substantial relationship amongst every observed determinant and the essential latent constructs indicates that the theory was accepted since all observed determinants had sig worth below 5% with loadings on factor more compared to 70%.

Table 2. Total variance explained

Co	Eigen-values			Squared Loadings			Rotation Sums		
	Sum	% of Variance	Cumulative %	Total	% of Variance	Cumulative %	Total	% of Variance	Cumulative %
1	15.46	28.634	28.634	15.463	28.63	28.634	5.622	10.411	10.411
2	4.709	8.720	37.355	4.709	8.720	37.355	5.377	9.957	20.368
3	2.489	4.609	41.964	2.489	4.609	41.964	3.727	6.903	27.271
4	2.363	4.375	46.339	2.363	4.375	46.339	3.507	6.494	33.765
5	2.062	3.819	50.158	2.062	3.819	50.158	3.397	6.291	40.055
6	1.944	3.600	53.758	1.944	3.600	53.758	3.013	5.579	45.634
7	1.737	3.217	56.975	1.737	3.217	56.975	2.596	4.808	50.442
8	1.416	2.623	59.598	1.416	2.623	59.598	2.543	4.710	55.152
9	1.391	2.575	62.173	1.391	2.575	62.173	2.363	4.377	59.529
10	1.208	2.237	64.410	1.208	2.237	64.410	2.172	4.023	63.552
11	1.133	2.098	66.509	1.133	2.098	66.509	1.597	2.957	66.509
12	.990	1.833	68.342	-	-	-	-	-	-

The trajectory of the correlation among each parameter, as depicted in Table 3, is measured by covariance. Positive covariance occurs when both determinants consistently show high or low values together. Sometimes, the determinants are low and one high, even if all covariance is positive. (Model 1: Group by default) To put it plainly, they were all founded significantly apart from zero.

Table 3. Covariance.

Covariance's			Estimate	Std. Error	Critical ratio	P
Investing-decision	<-->	Financing-decision	.199	.037	5.378	***
Investing-decision	<-->	Internal-Control	.109	.020	5.450	***
Investing-decision	<-->	Cost-Management	.249	.029	8.586	***
Investing-decision	<-->	Asset-Management	.179	.031	5.774	***
Financing-decision	<-->	Internal-Control	.107	.020	5.735	***
Financing-decision	<-->	Cost-Management	.259	.035	7.401	***
Financing-decision	<-->	Asset-Management	.241	.029	8.310	***
Internal-Control	<-->	Cost-Management	.102	.021	4.857	***
Internal-Control	<-->	Asset-Management	.051	.014	3.642	***
Cost-Management	<-->	Asset-Management	.241	.031	7.774	***

Model Fitness

The investigators observed that the CMIN ratio was below 3.0, as shown in Table 4, reflecting model approval. The Tucker-Lewis Index (TLI), which has an iterative conformance score of more than 0.90, was established by the investigators to contrast the model under test with the benchmark version. Researchers discovered that AI and human capital mediated the association between financial management techniques and business success. Financial management practices and corporate success were shown to be correlated in 18.2% of cases, but this has now increased to 43.5%.

Table 4. Mediation inquiry.

-	FM	HC	AI	OP
HC	1.671	.000	.000	.000
OP	.873	.075	.668	-------

A direct association of 0.866 was manifested between the firm's performance and FM strategies. This indicates that business performance is influenced by a plan to manage the finances by 87.3%. Additionally, Strategies to manage finances had a 167.1% influence on firm performance through AI and human capital. Performance gains of approximately 74.8% are attributed to the combination of AI and human capital mediation. Thus, a partial mediation between Coffee Cooperative Performance and Financial Management Strategies was established using AI and Human Capital. Therefore, partial mediation was depicted in the model.

CONCLUDING REMARK

Higher levels of education, in the form of human capital and artificial intelligence, are frequently associated with greater social assets and economic success [32]. Human capital buildup, like physical capital accumulation, generates long-term potential returns on current investments. A direct correlation of 0.866 exists between coffee collaborative success and financial management strategies. This indicates that financial management approaches accounted for 86.6% of the coffee firms' performance. After human capital and artificial intelligence became readily accessible, fiscal oversight measures still had a 158.3% negative impact on the performance of coffee cooperatives. The combination of human capital and AI resulted in a roughly 75% increase in performance. As a result, we found a partial mediation link between financial management strategies and the efficiency of coffee cooperatives. Therefore, due to the influence of human capital and artificial intelligence, there is a partial mediation between coffee cooperative performance and financial management strategies.

MANAGERIAL AND PRACTICAL IMPLICATIONS

Study outcomes manifest that Ethiopian coffee cooperatives should prioritize enhancing the human assets of their staff. This encompasses several courses and training sessions designed to raise awareness of financial management. It is established that improving human capital is one way to enhance the overall financial performance of cooperatives. Coffee cooperatives should consider utilizing Artificial Intelligence (AI) applications for financial management. AI-driven forecasting, analysis, and decision-making will enable cooperatives to make more informed and nuanced financial decisions. Thus, AI technology may help improve performance, and economic effectiveness and productivity could also be enhanced. Cooperatives should encourage pilot solutions and innovative mindsets in a way that enables them to adopt new possibilities provided by AI solutions. Managers should promote innovation among staff members and those involved in the organization to develop better ideas and potentially implement more advanced technology. This will ensure that cooperatives implement AI technologies effectively in managing their finances, undertake the process efficiently, and keep them informed about the latest advancements. Entities that deal in the sales of coffee should foster staff members' and members' cooperation and information sharing. This can be done through training sessions, workshops, and meetings with key stakeholders. Cooperatives can enhance performance and effectively operate in challenging situations by sharing the issues of proper Governance and managing funds and knowledge on AI applications.

Coffee cooperatives should design and organize courses that will address the needs of their employees and members in managing their financial resources. Risk management, financial analysis, and budgeting should be part of these courses. This is achieved through the influence of a setting that tailors the training program to the needs of coffee cooperatives, thereby equipping them with skills that have immediate and positive impacts on financial returns. Coffee cooperatives may wish to explore partnerships with AI sellers or IT organizations that specialize in financial management. Some of these may help improve the ease and manner in which AI applications can be integrated into cooperative operations. Special attention should be paid to selecting reliable partners who understand the specific challenges faced by Ethiopian coffee cooperatives. Coffee cooperatives should establish bodies to oversee the effectiveness of the AI applications and financial management tools they implement. The following are examples of how one could accomplish this: financial audits, employee or member communication channels, and KPIs based on key indicators. Coffee cooperatives should also be eligible for regional, national, and international knowledge exchange platforms. This may include meetings for conferences, seminars, and online forums related to artificial intelligence and inefficiency in financial management strategies in the agricultural sector. Through these avenues, Cooperatives can access current research, trends, and even learn from the successes of others.

AUTHORS' CONTRIBUTIONS

Mershaye Birhane was involved in the section on interpretation and analysis. Based on a theoretical framework and a review of the literature, Dr. Chalchisa Amentie contributed to the development of the hypothetical framework. Professor Brehanu Borji provided the framework for the research technique. Mershaye Birhane and Professor Brehanu Borji collaborated to establish the sampling methodologies and processes. Dr. Shashi Kant gained access to the model fitness. Mershaye Birhane would thus like to express his gratitude to Professor Brehanu Borji and Associate Professor Chalchisa Amentie, his supervisors, for their invaluable assistance during the writing and critical feedback stages. This is a portion of Mershaye Birhae's doctoral study. This work is part of PhD work of Mershaye Birhae's PhD work.

REFERENCES

[1] D. Mhlanga, "Industry 4.0 in finance: the impact of artificial intelligence (AI) on digital financial inclusion", *International Journal of Financial Studies,* vol. 8, no. 3, p. 45, 2020. [http://dx.doi.org/10.3390/ijfs8030045]

[2] Oluwafunmilola Oriji, Oluwabosoye Abitoye, Chibuike Daraojimba, O. Abitoye, and C. Daraojimba, "Financial technology evolution in Africa: a comprehensive review of legal frameworks and implications for ai-driven financial services", *International Journal of Management & Entrepreneurship Research,* vol. 5, no. 12, pp. 929-951, 2023. [http://dx.doi.org/10.51594/ijmer.v5i12.627]

[3] G. Anwar, and N.N. Abdullah, "The impact of Human resource management practice on Organizational performance", *Int. J. Eng. Bus. Manag.,* p. 5, 2021. [IJEBM].

[4] M.D. Abdi, H.A. Dobamo, and K.B. Bayu, "Exploring current opportunity and threats of artificial intelligence on small and medium enterprises accounting function; evidence from south west part of ethiopia, oromiya, jimma and snnpr, bonga", *Academy of Accounting and Financial Studies Journal,* vol. 25, no. 2, pp. 1-11, 2021.

[5] V. Gaikar, and J. Cherian, "An empirical study on impact of sustainability-oriented total quality management strategies on organizational performance in UAE based organization. In An Empirical Study on Impact of Sustainability-Oriented Total Quality Management Strategies on Organizational Performance in UAE Based Organization: Gaikar, Vilas| uCherian, Jacob", *SSRN,* 2021. [Sl]

[6] "", , 2020pp. 1-6 S. M. Anicet Kiemde and A. D. Kora, "The challenges facing the development of AI in Africa", *2020 IEEE International Conference on Advent Trends in Multidisciplinary Research and Innovation (ICATMRI), 2020pp. 1-6.*

[7] N.R. Mosteanu, and A. Faccia, "Digital systems and new challenges of financial management–FinTech, XBRL, blockchain and cryptocurrencies", *Qual. Access Success,* vol. 21, no. 174, pp. 159-166, 2020.

[8] A.K. Sampene, F.O. Agyeman, B. Robert, and J. Wiredu, "Artificial intelligence as a path way to Africa's transformations", *Artif. Intell.*, vol. 9, no. 1, 2022.

[9] B. George, R.M. Walker, and J. Monster, "Does strategic planning improve organizational performance? A meta-analysis", *Public Adm. Rev.*, vol. 79, no. 6, pp. 810-819, 2019. [http://dx.doi.org/10.1111/puar.13104]

[10] S.A. Finkler, T.D. Calabrese, and D.L. Smith, *Financial management for public, health, and not-for-profit organizations.* CQ Press, 2022.

[11] A. Chilunjika, K. Intauno, and S.R. Chilunjika, "Artificial intelligence and public sector human resource management in South Africa: Opportunities, challenges and prospects", *SA J. Hum. Resour. Manag.*, vol. 20, p. 12, 2022. [http://dx.doi.org/10.4102/sajhrm.v20i0.1972]

[12] E. Nizam, A. Ng, G. Dewandaru, R. Nagayev, and M.A. Nkoba, "The impact of social and environmental sustainability on financial performance: A global analysis of the banking sector", *J. Multinat. Financ. Manag.*, vol. 49, pp. 35-53, 2019. [http://dx.doi.org/10.1016/j.mulfin.2019.01.002]

[13] E.F. Brigham, and J.F. Houston, *Fundamentals of financial management: Concise.* Cengage Learning, 2021.

[14] F. Bartolacci, A. Caputo, and M. Soverchia, "Sustainability and financial performance of small and medium sized enterprises: A bibliometric and systematic literature review", *Bus. Strategy Environ.*, vol. 29, no. 3, pp. 1297-1309, 2020. [http://dx.doi.org/10.1002/bse.2434]

[15] P. Giudici, "Fintech risk management: A research challenge for artificial intelligence in finance", *Front. Artif. Intell.*, vol. 1, p. 1, 2018. [http://dx.doi.org/10.3389/frai.2018.00001] [PMID: 33733089]

[16] C. Baah, D. Opoku-Agyeman, I.S.K. Acquah, Y. Agyabeng-Mensah, E. Afum, D. Faibil, and F.A.M. Abdoulaye, "Examining the correlations between stakeholder pressures, green production practices, firm reputation, environmental and financial performance: Evidence from manufacturing SMEs", *Sustain. Prod. Consum.*, vol. 27, pp. 100-114, 2021. [http://dx.doi.org/10.1016/j.spc.2020.10.015]

[17] D.N. Dwivedi, G. Mahanty, and Y.K. Pathak, "AI applications for financial risk management", In: *The impact of AI innovation on financial sectors in the era of industry 5.0* IGI Global, 2023, pp. 17-31. [http://dx.doi.org/10.4018/979-8-3693-0082-4.ch002]

[18] N.H. Tien, D.B.H. Anh, and N.M. Ngoc, "Corporate financial performance due to sustainable development in Vietnam", *Corp. Soc. Resp. Environ. Manag.*, vol. 27, no. 2, pp. 694-705, 2020. [http://dx.doi.org/10.1002/csr.1836]

[19] M. Javed, M.A. Rashid, G. Hussain, and H.Y. Ali, "The effects of corporate social responsibility on corporate reputation and firm financial performance: Moderating role of responsible leadership", *Corp. Soc. Resp. Environ. Manag.*, vol. 27, no. 3, pp. 1395-1409, 2020. [http://dx.doi.org/10.1002/csr.1892]

[20] H. Ndoro, K. Johnston, and L.F. Seymour, "Artificial intelligence uses, benefits and challenges: a study in the western cape of South Africa financial services industry", *SACAIR,* vol. 2020, p. 58, 2020.

[21] Y. Zhang, U. Khan, S. Lee, and M. Salik, "The influence of management innovation and technological innovation on organization performance. A mediating role of sustainability", *Sustainability (Basel),* vol. 11, no. 2, p. 495, 2019. [http://dx.doi.org/10.3390/su11020495]

[22] H.A. Meressa, "Growth of micro and small scale enterprises and its driving factors: empirical evidence from entrepreneurs in emerging region of Ethiopia", *J. Innov. Entrep.*, vol. 9, no. 1, p. 11, 2020. [http://dx.doi.org/10.1186/s13731-020-00121-9]

[23] V.B. Arsic, "Challenges of financial risk management: AI applications", *Management: Journal of Sustainable Business and Management Solutions in Emerging Economies,* vol. 26, no. 3, pp. 27-34, 2021.

[24] S. Addis, "An exploration of quality management practices in the manufacturing industry of Ethiopia", *TQM J.,* vol. 32, no. 1, pp. 127-142, 2019. [http://dx.doi.org/10.1108/TQM-01-2019-0031]

[25] S. Oduro, and L.G. Haylemariam, "Market orientation, CSR and financial and marketing performance in manufacturing firms in Ghana and Ethiopia. Sustainability Accounting", *Management and Policy Journal,* vol. 10, no. 3, pp. 398-426, 2019.

[26] M. Hidayat, S.Y. Defitri, and H. Hilman, "The Impact of Artificial Intelligence (AI) on Financial Management", *Management Studies and Business Journal (PRODUCTIVITY),* vol. 1, no. 1, pp. 123-129, 2024. [PRODUCTIVITY]. [http://dx.doi.org/10.62207/s298rx18]

[27] O. Melnychenko, "Is artificial intelligence ready to assess an enterprise's financial security?", *J. Risk Financ. Manag.,* vol. 13, no. 9, p. 191, 2020. [http://dx.doi.org/10.3390/jrfm13090191]

[28] P. Polak, C. Nelischer, H. Guo, and D.C. Robertson, "“Intelligent” finance and treasury management: what we can expect", *AI Soc.,* vol. 35, no. 3, pp. 715-726, 2020. [http://dx.doi.org/10.1007/s00146-019-00919-6]

[29] S. Fritz-Morgenthal, B. Hein, and J. Papenbrock, "Financial risk management and explainable, trustworthy, responsible AI", *Front. Artif. Intell.,* vol. 5, p. 779799, 2022. [http://dx.doi.org/10.3389/frai.2022.779799] [PMID: 35295866]

[30] X. Zheng, M. Zhu, Q. Li, C. Chen, and Y. Tan, "FinBrain: when finance meets AI 2.0", *Frontiers of Information Technology & Electronic Engineering,* vol. 20, no. 7, pp. 914-924, 2019. [http://dx.doi.org/10.1631/FITEE.1700822]

[31] E. O. Arakpogun, Z. Elsahn, F. Olan, and F. Elsahn, "Artificial intelligence in Africa: Challenges and opportunities", *The fourth industrial revolution: Implementation of artificial intelligence for growing business success,* pp. 375-388, 2021.

[32] P. N. Kenyanya, and D. Ndegwa, "Financial factors, financial literacy and coffee output among smallholder coffee farmers in Kenya", *Eur J Bus Manag Res,* vol. 9, no. 6, pp. 52-56, 2024. [http://dx.doi.org/10.24018/ejbmr.2024.9.6.2488]

[33] S. Kant, and M. Adula, Human-Machine Interaction in the Metaverse in the Context of Ethiopia.*Impact and Potential of Machine Learning in the Metaverse.* IGI Global, 2024, pp. 196-212. [http://dx.doi.org/10.4018/979-8-3693-5762-0.ch008]

[34] K. Guyo, and C.A. Kero, "Mediation of Marketing Intermediaries and AI Adoption Between Livestock Products Marketing and Economic Status of Pastoralist in Ethiopia", *2023 International Conference on Communication, Security and Artificial Intelligence (ICCSAI),* 2023pp. 311-316 Greater Noida, India [http://dx.doi.org/10.1109/ICCSAI59793.2023.10421261]

[35] S. Kant, and M. Adula, Mediated by AI-Based Generative Re-Enforcement Learning and Work Attitude: Are Intrinsic Rewards Transforming Employee Perceived Organizational Support?*Generative AI for Transformational Management.,* J. Gomathi Sankar, A. David, Eds., IGI Global, 2024, pp. 83-108. [http://dx.doi.org/10.4018/979-8-3693-5578-7.ch004]

[36] K. Asefa, and K.L. Debela, "Effect of Transformational Leadership on Organizational Performance: The Mediating Role of Employee Commitment and Expert Systems (AI) Inclusion in Ethiopia", *2023 International Conference on Communication, Security and Artificial Intelligence (ICCSAI),* 2023pp. 305-310 Greater Noida, India

[http://dx.doi.org/10.1109/ICCSAI59793.2023.10421269]

[37] S. Kant, and M. Adula, AI Learning and Work Attitude Mediation Between Reward and Organizational Support in Ethiopia.*Generative AI for Transformational Management.,* J. Gomathi Sankar, A. David, Eds., IGI Global, 2024, pp. 109-136.
[http://dx.doi.org/10.4018/979-8-3693-5578-7.ch005]

[38] M. Birhane, C. Amentie, B. Borji, and S. Kant, Are Ethiopian Coffee Farmers Proficient to Predict Capital Structure with Mediation of AI Based Knowledge Systems and Asset Liquidity?
[http://dx.doi.org/10.1201/9781003451648-15]

[39] S. Potluri, S. Satpathy, S.S. Basa, A. Zuorro, Ed., *AI in Agriculture for Sustainable and Economic Management.* CRC Press, 2024.
[http://dx.doi.org/10.1201/9781003451648]

Section 2
Intelligent AI & ML Solutions

CHAPTER 8

Smart Cities: Urban Planning and Infrastructure Management

Aanchal Chaudhary[1], **Harsh Kumar Pandey**[1], **Ayush Yadav**[1], **Vaibhav Kumar Singh**[1], **Nishant Singh**[1] and **Hitesh Mohapatra**[1,*]

[1] *School of Computer Engineering, KIIT (Deemed to Be) University, Bhubaneswar-751024, Odisha, India*

Abstract: The Internet of Things (IoT) can access certain subsets of data for the creation of several digital services, while also integrating a wide range of diverse and heterogeneous end systems in a transparent and seamless manner. This chapter examines innovative solutions to key challenges in smart cities, with an emphasis on ambient noise assessment, air quality monitoring (E-nose system), and smart parking management systems (SPMS). Utilizing servo motors, smartphone apps, and sensors, the SPMS combines Internet of Things technology to maximize parking space distribution, improve user experience, and raise total parking efficiency. An E-nose air quality monitoring system with a sensor array based on the ESP32 can identify gas pollutants, dust, and CO_2. Techniques for processing data, such as threshold alarms and median computation, provide precise and prompt air quality monitoring. Additionally, a noise monitoring system with a Gaussian filter and sound level sensors is demonstrated. The proposed analysis demonstrates how IoT-based solutions may be used to enhance sustainability, manage urban infrastructure more effectively, and improve people's quality of life in smart cities.

Keywords: Internet of Things, Internet of Things network, Monitoring, Smart cities, Urban areas.

INTRODUCTION

The significant development in population density within urban areas has necessitated the provision of substructures and services to fulfill the needs of inhabitants. Since it is now possible to connect devices and enable communication between them *via* the Internet, new smart city solutions are emerging to address these needs.

* **Corresponding author Hitesh Mohapatra:** School of Computer Engineering, KIIT (Deemed to Be) University, Bhubaneswar -751024, Odisha, India; hiteshmahapatra@gmail.com

Suneeta Satpathy, Sachi Nandan Mohanty & Subhendu Kumar Pani (Eds.)

There has been a notable increase in the number and quality of digital devices, including sensors, actuators, smartphones, and smart appliances, which supports the broad commercial objectives of the Internet of Things (IoT) [1]. Thus, the Internet of Things idea promises to make the Internet even more immersive and ubiquitous. Additionally, the Internet of Things (IoT) will facilitate the creation and development of several applications that utilize the potentially massive volume and diversity of data generated by extensive range of devices, such as household appliances, security cameras, monitoring sensors, actuators, vehicles, and many more by facilitating simple access and interaction and thus offering new services to individuals, businesses, and government agencies for public administration [2]. Fig. (**1**) shows some IoT devices that are used to make cities smart.

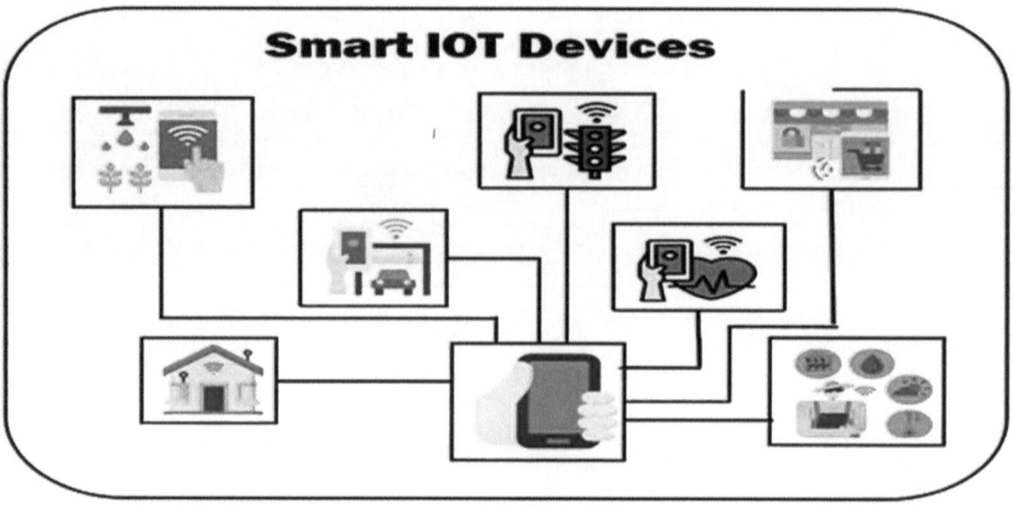

Fig. (1). Smart IoT devices.

It is a difficult task to find solutions that can meet the needs of every potential application scenario in such a diverse field of application. Numerous and, perhaps, incompatible proposals for the practical implementation of IoT systems have proliferated as a result of this difficulty [3]. Due to its novelty and complexity, the implementation of an Internet of Things network, along with the necessary backend network services and devices, still lacks an established best practice from a systems perspective. A well-defined and widely accepted business model that can draw investments to support the deployment of these technologies is lacking, which hinders the adoption of the IoT paradigm, in addition to its technical challenges [4].

Waste Management

Waste management poses a significant barrier for smart city applications, affecting both the environment and the health of our society. Humans and animals produce solid waste, which is usually discarded as useless [5]. At least 33% of the 2.01 billion tons of municipal solid waste produced worldwide each year is not managed in an environmentally responsible manner. By 2050, global waste is estimated to reach 3.40 billion tons, more than doubling population growth over that period [6].

Air Quality

The World Health Organization (WHO) has released a report on air quality that highlights the significant environmental threat that air pollution poses to human health. The research reported nearly 6 million premature deaths as a result of exposure to contaminated air sources [7]. Data from monitoring the busiest metropolitan regions demonstrate that pollution affects not only our own lives but also the lives of future generations. Consequently, it is critical that we make every effort to exercise caution and regulate the hazardous pollutants that we emit into the environment [8]. Installing pollution and air quality sensors throughout the city and making the sensor data accessible to the general public are necessary for the implementation of such a service [9].

Noise Monitoring

Chronic noncommunicable diseases are largely caused by classical risk factors like smoking, diabetes, and arterial hypertension; however, more recent research has indicated that environmental variables also play a role in the development of these diseases. In our industrialized society, environmental stressors, such as air pollution and noise, are becoming increasingly significant. In particular, traffic noise from automobile, airplane, and train travel may be a new risk factor for cardiovascular diseases, such as hypertension, hyperglycemia, hyperlipidemia, and increased blood viscosity and coagulation [10]. In order to measure the amount of noise generated in the locations with these services, an urban IoT can provide a noise monitoring service.

Smart Parking

India is the second most populous country in the world. The growth and development of Indian cities have resulted in a lack of appropriate parking space architecture. As individuals frequently travel for work, there is a high possibility that they may not be familiar with the neighborhood. As a result, they might not be aware of where to park [11]. Road sensors and intelligent displays, which show drivers the best parking route in the city, are the foundation of the smart parking service.

City Energy Consumption

In addition to the many benefits that smart cities offer, some challenges need to be overcome, including environmental quality and energy resource optimization. One of the main issues smart cities face is energy optimization, which includes managing resources, promoting active energy efficiency, and achieving optimal energy utilization [12]. An urban Internet of Things (IoT) could offer energy consumption monitoring for the entire city. This would provide authorities and the public with a clear understanding of the energy requirements for various services, such as traffic lights, public lighting, heating and cooling of public buildings, and surveillance cameras [13].

LITERATURE REVIEW

Due to their rapid growth, modern cities have been dealing with difficulties that affect urban quality of life, requiring improvements in areas, such as environmental monitoring, public services, and managerial efficiency. Smart city technologies seem to offer numerous feasible solutions to these and other contemporary urban difficulties [14]. A smart city is a contemporary metropolis that has been intelligently and sustainably functionalized to guarantee efficiency and sustainability. To accomplish this, it will be necessary to integrate various infrastructures and services into coherent units that can be monitored and managed by smart devices. The primary goals of smart cities are to address crime, education, energy, the environment, healthcare, public transportation, job creation, waste management, and the strategic and shared use of buildings and other city spaces, as well as the management of automobiles and even pets. Different urban information systems, which include networking support structures and sensor-level data, can be constructed using cloud and IoT-based data management [15]. Numerous studies have examined the advantages and primary challenges of IoT technology in smart cities. Within this category, the most extensively researched issues are smart mobility and transportation based on IoT systems. "Smart Velomobility", roadmap framework, smart parking, and IoT-enabled multi-agent systems with sensors dispersed throughout roadways are a few examples. Additionally, security issues and sensing technologies have an impact on how the Internet of Things is used in smart cities. Additionally, the integration of human stakeholders and physical infrastructure in smart cities has received special attention recently.

Over the past ten years, the academic and practice communities have shown a great deal of interest in the Internet of Things. Specific operating systems and communication protocols are required to enable user-device interactions in order to achieve core IoT features [16]. Fig. (**2**) provides a wide picture of the Internet of Things in the application sector. Advancements in information and communications technology (ICT) have made modern cities "smarter" and more effective in many ways. Nevertheless, it is not feasible or even required to make every element of smart cities "smart". The deployment of IoT in various application domains has been the focus of several European projects funded by the 7th Framework Program for Research and Technological Development (FP7), such as Advancing Identification Matters (AIM), District Information Modeling and Management for Energy Reduction (DIMMER), Intelligent Use of Building Energy Information (IntUBE), and Smart Energy Efficient Middleware for Public Spaces (SEEMPubS). To date, communication and network technologies have been utilized in smart cities to address contemporary urban challenges, such as population growth, overcrowding, and traffic congestion [17].

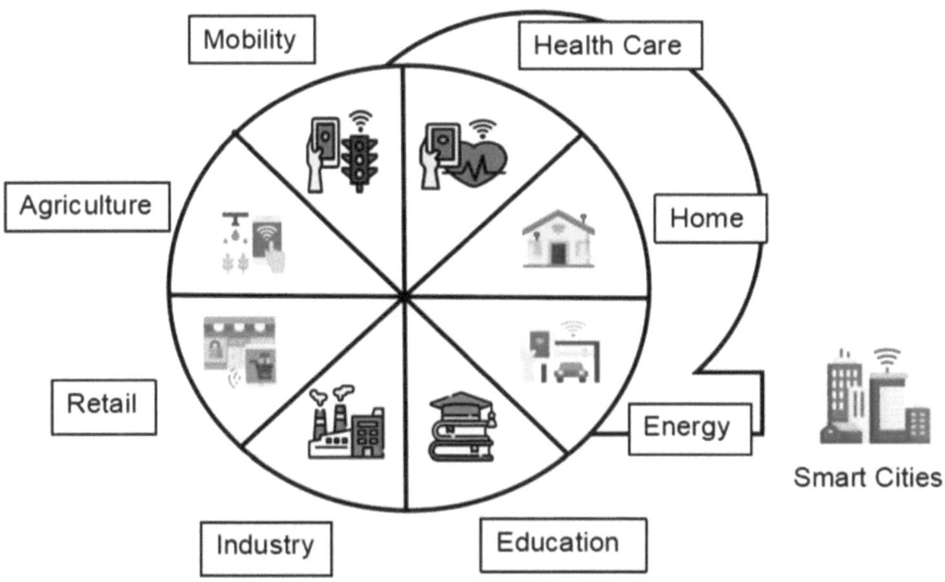

Fig. (2). IoT in different sectors.

The cost of waste collection can be decreased and the quality of recycling can be improved by using intelligent waste containers, which can detect the amount of waste and optimize the route of the collector truck. The Internet of Things will connect end devices, such as intelligent waste containers, to a control center, where data is processed by optimization software to determine the most efficient way to manage the fleet of collection trucks [18]. According to the UN

Environment program, the Paris Agreement was signed in 2015. It was estimated that, due to existing policies, greenhouse gas emissions would increase by 16 percent by 2030. The anticipated rise for now is 3%. However, to follow the 2°C Paris Agreement and the 1.5°C pathway, estimated 2030 greenhouse gas emissions still need to decrease by 28% and 42%, respectively. Fitness trails, parks, and crowded places can all have their air quality monitored by an urban IoT. In addition to being able to consistently access their preferred personal training application, people can always find the healthiest route for outdoor activities. Sensors for pollution and air quality must be installed throughout the city, and citizen access to the sensor data is necessary for the implementation of such a service.

Through the use of sound detection algorithms that can identify sounds like glass breaking or fights, a noise monitoring service can be used to enforce public safety in addition to creating a space-time map of the noise pollution in the area. The installation of sound detectors or environmental microphones is highly contentious due to the evident privacy concerns associated with this type of monitoring; however, this service can also enhance both the calm of city nights and the confidence of public establishment owners [19].

The urban IoT infrastructure can seamlessly incorporate smart parking services, as numerous European companies offer products specifically designed for this purpose. Additionally, an electronic parking permit verification system can be implemented in designated spots for residents or the disabled using short-range communication technologies such as Near Field Communication (NFC) or Radio Frequency Identification (RFID). This will enhance services for citizens who are legally entitled to occupy these spaces and provide an effective means of swiftly identifying infractions. According to the World Energies and Climate Statistics for December 2023, natural gas consumption increased by 9.5% compared to the same period in 2022, reaching nearly 38 billion cubic meters (bcm). Power draw monitoring devices must be integrated with the city's power grid to receive such a service, as presented in Table **1**. Furthermore, these services can be improved by adding active functionalities that regulate local power production structures [20].

Table 1. Services in smart city

Services	**Network**	**Sensors Used**
Waste Management	WiFi: 4G, 5G, and 6G	Ultrasonic, robust, water, disposal sensor, and shock resistant
Air Quality	Bluetooth and WiFi	Airly PM+GAS type 1, Airly PM+GAS type 2
Noise Monitoring	Bluetooth, Wi-Fi, and NFC	Environmental sensors, acoustic sensors, sound level meters (SLMs)

(Table 1) cont.....

Services	Network	Sensors Used
Smart Parking	Ethernet and Wi-Fi	ultrasonic sensors, infrared sensors
City Energy Consumption	PLC and Ethernet	electricity meters, gas meters, water meters

RESEARCH GAP

The Internet of Things (IoT) holds the promise of digitalizing nearly every aspect of our lives. The process of digitization for smart cities involves the widespread installation of sensing nodes across all areas of the city's operational framework. There are many obstacles to be taken into account when developing and implementing IoT systems in smart cities, especially with such a wide range of applications. This section covers the difficulties that IoT system designers encounter when deploying their products in smart city applications. The technological issues that have drawn the attention of researchers and are relevant to the use of IoT in smart cities are the main topic of this chapter [21].

Security and Privacy

In smart cities, security and privacy are of utmost importance. Any disruption in city service operations can cause inconvenience to residents and pose a threat to their safety and property. Smart cities must have all of their vital infrastructures online. Thus, in smart cities, security is a major concern. Today's smart cities are more vulnerable than ever to cybercrime and warfare, which have become common political strategies worldwide. In this case, data being transferred across the network must be encrypted. Citizens' involvement and trust are essential for smart city initiatives to succeed.

Smart Sensors

The hardware elements of smart cities' data collection systems are called smart sensors. Numerous vendors produce these devices, each adhering to different data formats, connectivity protocols, measurement standards, and sensing mechanisms. When smart cities are deployed, all of these devices will need to be able to exchange data, schedule tasks amongst themselves, and aggregate data to draw conclusions. Robustness and dependability present another difficulty for smart sensors. The dependability and accuracy of the IoT system are referred to as robustness and reliability. IoT is the backbone of future smart cities, and since it is essential to their operation, the system must offer its users a seamless experience. This necessitates that users of the application receive accurate and prompt responses to their service requests. Every inhabitant of the smart city must receive high-quality services [22].

Networking

The ability of sensors and other devices to transmit and receive data to and from the Cloud is essential to the Internet of Things. A major challenge is enabling networking for these devices to stay connected as new applications for smart cities emerge. The networking techniques used today are not suited to offer networking services for the various components of smart cities. In order to deliver a service of a reasonable caliber, many smart city devices have mobility and data throughput requirements that must be satisfied. Various methods have been proposed to define local networks, access points, and other related entities in order to resolve this issue. Working on effective and dynamic routing protocols is another facet of networking that can meet the requirements of the Internet of Things by supporting both stationary and mobile devices, a capability that many existing protocols do not adequately provide [23].

Big Data Analytics

The core technology of smart city systems is sensing. The information and data used to create innovations for smart cities are collected by sensors. Many different types of sensors are employed in smart city projects due to their diverse components and varied nature. In addition to listing the sensors used in the Internet of Things, a framework is also provided for comparing them. The interaction between the smart city system and its residents, as well as the development of new services, is made possible by sensors, which are a crucial part of IoT systems for smart cities [24].

PROPOSED STUDY

Smart Parking Management Systems (SPMS) are a promising solution to one of the key issues in smart cities: the inefficient management of parking spaces. The primary goal of an SPMS is to provide drivers with a smart and effective means to quickly locate their parked cars, while also efficiently allocating available parking spots to them and reducing the time they spend searching for them [25]. To address some problems with parking structures, an Internet of Things (IoT) based E-parking system was created. The key features of the proposed system encompass several components. Based on the intervals between sending and receiving a signal, the parking loT monitoring component periodically assesses the condition of the parking spaces. A picture is captured and sent to the parking management system as soon as a change is noticed. A warning is given by the system if an automobile is parked incorrectly because it is unable to retrieve the license plate number. Reservations in this method are made *via* the mobile application that users download. Nevertheless, it lacks a practical means of finding empty parking spaces fast, which is a crucial function that is particularly required in huge multi-story parking lots.

Using sensors and a light coding method, an efficient parking solution was proposed that uses a light to indicate whether the space is occupied or free. A blue light indicates an occupied area in this scheme, a red light indicates an inaccurately filled spot, and a yellow light indicates an empty spot. This system differs from the others in that it provides two distinct parking options: business and economy. The primary distinction between them is their location within the parking garage and their cost; business spots are more expensive and closer to the exit. In multi-story parking lots, drivers may have to wait longer to reach the floor where the yellow light is on, which reduces the overall effectiveness of the parking system. However, using lights to indicate the status of parking spots may be useful if the vehicle needs to be parked on the same level (floor) as the available spot. Additionally, drivers need to be able to see the light connected to the open spot, which can be challenging in parking lots with bigger cars obstructing the view or pillars separating floors [26].

A servlet and SQL database are connected, enabling data retrieval and transmission, as shown in Fig. (**3**). Data delivered to the application/controller or *via* the Message Queuing Telemetry Transport (MQTT) protocol is used to control the servlet. This system relies heavily on the MQTT protocol, which gathers and processes data as needed. When the necessary conditions are satisfied, this data can also be used to control servo motors that open and close the parking barrier. The information gathered from the vehicles and the parking lot infrastructure will be kept in a database that can be accessed *via* NGINX and JPA mapping.

Phidget sensors are used to regulate all parking spots and RFID detectors. These sensors are connected with Java to process changes and deliver the necessary information when changes are detected. The prototype's parking barriers are likewise managed by Phidget; they will move (open or close) in response to the controller receiving the proper data. Information will be displayed to the user *via* the mobile application based on their current location and condition within the parking lot. The user will be able to pay for parking, view spot details for their parked car, locate the closest available space within the parking lot, and enter and exit the lot using this mobile app.

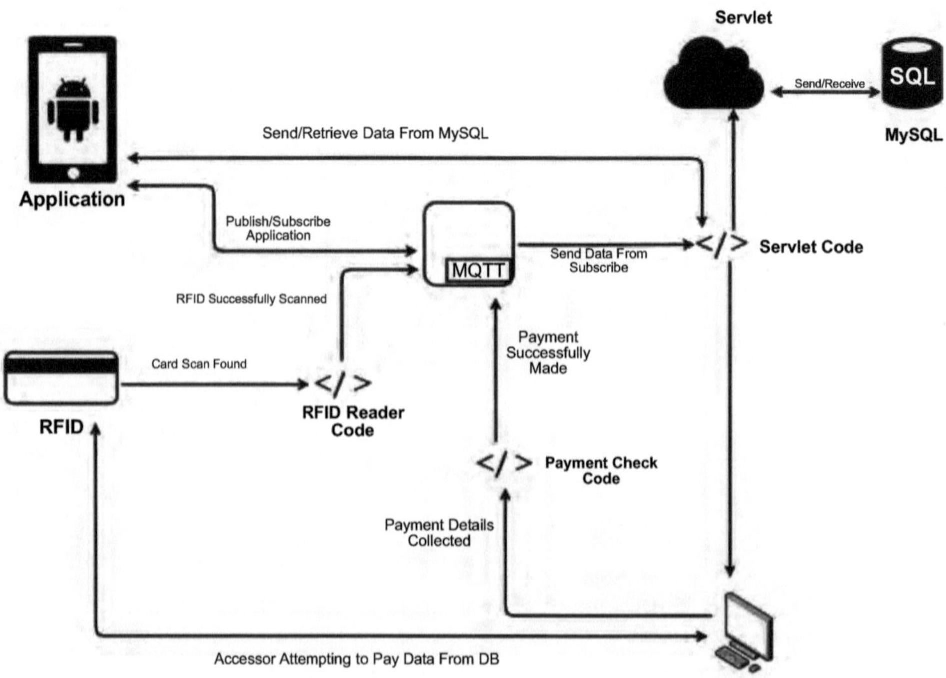

Fig. (3). Proposed SPMS architecture

In this chapter, an IoT-based e-nose air quality measurement system is also explored for real-time, affordable, and simple installation of air quality monitoring. It is a completely wireless solution developed using the ESP32 module, which integrates the IEEE 802.11 b/g/n network protocol into the IoT architecture. The two components of the e-nose air quality monitoring device are shown in Fig. (**4**). The first component is the detection and communication unit, which consists of an ESP32 microcontroller-based sensor array with integrated Wi-Fi. The second part is a mobile user interface based on iOS and Android [27].

The GP2Y1010AU is a dust sensor with an analogue output system. A phototransistor and an infrared light-emitting diode (IRED) are placed opposite one another. The phototransistor detects the infrared beams reflected by dust that enters the sensor's air chamber and produces a voltage in response [27]. The MH-Z14A CO_2 sensor module uses the non-dispersive infrared (NDIR) principle. It has a resolution of 5 ppm and measures between 0 and 5000 ppm, with an accuracy of ±50 ppm. Three separate output modes are provided by the sensor module for the CO_2 concentration: Pulse Width Modulation (PWM), analog output, and serial output (RS-232). The MICS-4514 is primarily used to measure emissions from vehicle exhausts, although it can also be used to measure gas concentrations, including hydrocarbons, NO_2, and CO. The sensor's upper side features a micro-sensing diaphragm and an integrated heating element. The MICS-4514 features two sensor chips with separate heaters and delicate layers. Oxidizing gases (OX) are detected by one sensor chip, while reducing gases (RED) are detected by the other sensor.

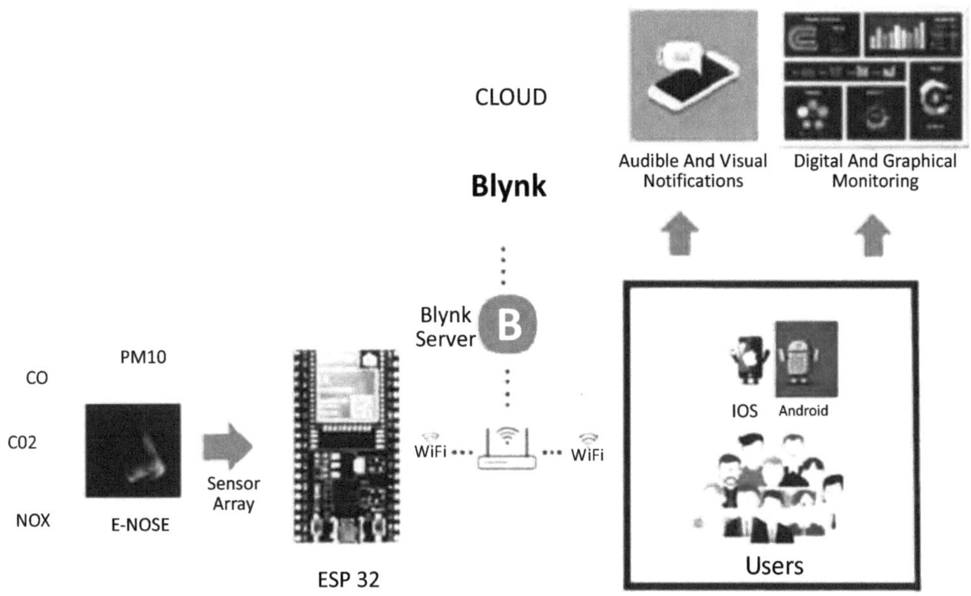

Fig. (4). Proposed E-nose system architecture.

The DHT22 consists of two parts: A thermistor temperature sensor and a capacitive humidity sensor. The DHT22 is a sophisticated sensor unit with a calibrated digital signal output. It features a quick response time and an 8-bit microcontroller. Its relative error for temperature measurement is ±0.5 °C, and for humidity measurement, it is ±2% rH. The dual-core ESP32 is equipped with numerous internal modules, including Bluetooth, Wi-Fi, RF, IR, CAN, Ethernet, temperature, hall effect, and touch sensors—all essential for smart home applications. The Harvard Tensilica Xtensa LX6 32-bit dual-core has a processor that can run at up to 240 MHz inside the ESP32 module framework. The sensors GP2Y1010AU, MH-Z14, MICS-4514, and DHT22, which measure temperature, humidity, CO_2, CO, PM10, and NO_2, are part of the detecting unit. In accordance with the measurement specifications, the sensors utilized in the system are installed in a sealed box measuring 17 × 12 × 8 cm. The box has a printed circuit board, a fan to provide airflow to the dust sensor, and a power supply to supply the system's energy in addition to the measuring sensors.

The median of 12 observations, taken at 5-second intervals, is used to calculate the climatic parameters and gas concentrations. This reduces the impact of inaccurate measurements resulting from malfunctioning sensors. When the specified gas concentration and climate parameter thresholds are exceeded, users receive an alert. Users are not sent misleading warnings; instead, alerts containing five-minute averages are computed and sent. There is no memory element employed in this e-nose air quality measurement system to record data. Through the mobile interface, the received data is directly recorded to the Blynk cloud server. When requested, the user's registered email address receives the data. Prior to every data transmission, the internet connection's status is examined. In order to maintain data integrity, data packets that cannot be temporarily communicated while connected are retained. Upon reestablishing the internet connection, these packets are subsequently forwarded to the cloud server using past time tags. By maintaining the data flow, data loss can be avoided in this way.

For iOS and Android smartphones, there are numerous free options. One such program is Blynk, an Internet of Things platform designed for iOS and Android apps that allows handling several controllers, including the Raspberry Pi, ESP8266, ESP32, chipKIT, Intel, LeMarker, Onion Omega, SparkFun, and STM32. Digital data, including temperature, humidity, current, voltage readings, and control systems, is stored and conveniently accessible at any time *via* the Blynk cloud server service. It is possible to use real-time clock and calendar (RTCC) functions with the Blynk graphical components, or widgets. This section illustrates the creation of a monitoring system that utilizes sensors to measure ambient noise.

The data is sent to a specific platform *via* the IoT paradigm and a Gaussian filter implemented by the system. The suggested architecture is integrated in two stages: data acquisition and preprocessing and data reception and IoT. The constituent parts of each phase are depicted in Fig. (**5**), which is explained in more detail below [28]. Two sound level sensors attached to an Arduino Nano development board are used by the monitoring system to collect and preprocess data. This stage also includes data storage and pre-processing, which includes removing noise from the data and smoothing the sensor signals using a Gaussian filter on the same board. Data are then transmitted *via* a radio transceiver to the data receive phase [29]. The DFRobot Gravity Analog Sound Meter (SEN0232) is utilized in the system; it is a calibrated sound level meter with a frequency response of 31.5 Hz to 8.5 kHz and a measuring range of 30 to 130dBA. It features an analog output with a voltage range of 0.6 to 2.6V and a time response of 125 ms.

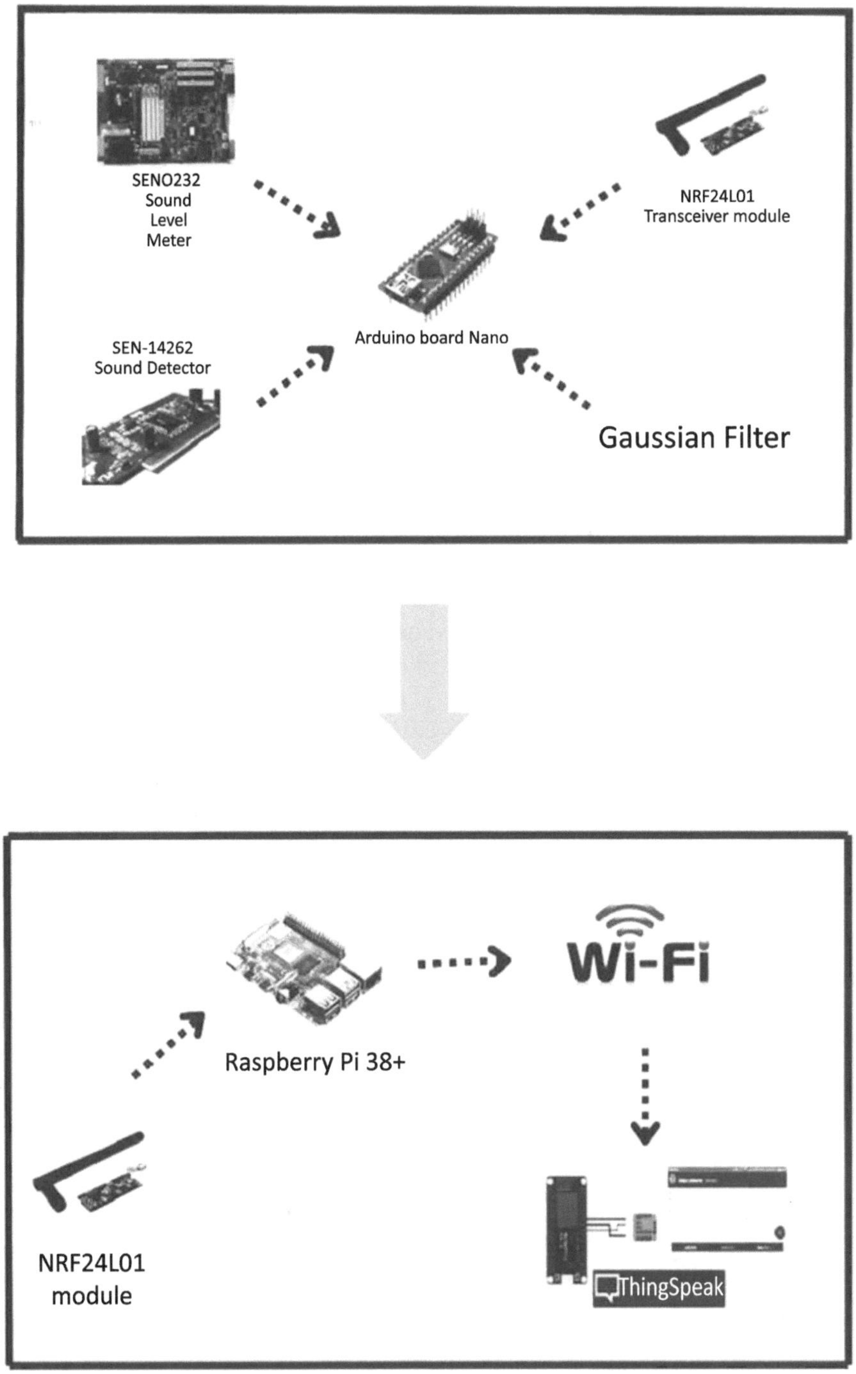

Fig. (5). Noise monitoring system architecture.

An additional Sparkfun Sound detector (SEN-14262) is also included in the design. Three separate outputs, an analog representation of the sound wave, an analog representation of its amplitude, and a binary representation of sound presence, are used by this sensor to determine whether sound is present. A Gaussian filter is also incorporated into this architecture with the aim of removing noise from the sensor signals. Using a Gaussian library, the filter is implemented using the method described previously [30].

In this chapter, another noise monitoring system is also explored, which is known as the Arduino Nano board. The Arduino Nano board utilizes the radio frequency module NRF24L01+ to transmit data. With an energy consumption of less than 115 mA, the NRF24L01+ enables data transmission across a 2.4 GHz ISM open band at data rates of 250 kbps, 1 Mbps, and 2 Mbps. It enables multiple operations between various devices, supporting 125 frequency points and six channels of data reception. The module features an SMA antenna with an operational range of up to 1100 meters, a Power Amplifier (PA), and a Low-Noise Amplifier (LNA). Furthermore, an RGB LED integrated into the prototype serves as a noise indicator. The intensity of the noise determines the color of the LED. The data collected is compared across scales from 30 dBA to 140 dBA, and the RGB LED alerts by emitting a color based on the sound strength.

A Raspberry Pi 3B+ is used for the data reception phase, incorporating the previously described radio frequency module receiver. Each data packet from the prototype is received and deserialized in JSON format during this step. The data is then transferred to the IoT platform and stored in a local database. Moreover, MongoDB is also employed as a Raspberry Pi-based local database service. MongoDB is a document-oriented, NoSQL database that uses a JSON-like format to store data. Consequently, there is no need to perform any further processing while storing deserialized data. To evaluate the collected data, the IBM ThingSpeak platform is employed; this cloud-based IoT analytics platform enables data aggregation, visualization, and analysis. We can exchange sensor data through information-displaying real-time channels on this platform. Channels may be public or private. Furthermore, a wireless connection and the MQTT protocol are used to send data every 30 seconds.

The proposed IoT-based solutions to inefficiencies in parking management within smart cities suggest an architecture for the Smart Parking Management System (SPMS) that combines several cutting-edge elements. The system utilizes Internet of Things (IoT) technologies, including Phidget sensors, RFID detectors, and mobile applications, to provide drivers with an efficient way to locate parking spots, schedule reservations, and navigate parking lots more effectively. The system is further automated by the addition of servo motors for parking barrier control, which improves the user experience. The SPMS is notable for its use of the MQTT protocol for data transmission and retrieval, ensuring smooth communication between various system components. This enables real-time monitoring of parking spots, allowing for quick responses to changes in parking availability. Moreover, reliable data management and accessibility for users and administrators are guaranteed by the integration of servlets, SQL databases, NGINX, and JPA mapping. For added flexibility and convenience, businesses and economy spots have different pricing options. Light-coding techniques are also included to show the availability of parking spaces. For the system to function optimally, however, obstacles, such as visibility issues in multi-story parking lots and potential delays in accessing available spots on higher floors, must be taken into account.

Beyond parking management, this chapter introduces an IoT e-nose air quality measurement system. The system utilizes various sensors, including a dust sensor, CO_2 sensor, gas concentration sensor, temperature sensor, and humidity sensor, all integrated into an ESP32 module to provide comprehensive environmental monitoring. Real-time data processing, combined with cloud storage and analysis *via* platforms, such as Blynk or IBM ThingSpeak, enhances the system's ability to provide users with real-time air quality data.

CONCLUSION

IoT-based solutions are designed to solve a range of issues in smart city infrastructures. One of them is a smart parking management system (SPMS). The SPMS is designed to manage parking spaces more effectively by leveraging IoT technologies like sensors, mobile apps, and the Mobile Quality Transfer Traffic (MQTT) protocol. The goal of the SPMS is to help drivers find parking spaces more quickly and allocate them more efficiently. Another framework discussed is an E-nose quality estimation framework that provides real-time air quality monitoring, using sensors, such as GP2Y1010AU, MH-Z14A, MICS-4514, and DHT22. This framework utilizes the ESP32 module and a portable client interface for collecting and visualizing data. It offers a remote and cost-effective solution for monitoring CO_2 levels, particulate matter, and other gas concentrations. Lastly, this chapter introduces a noise monitoring system that utilizes sound level sensors, an Arduino Nano board, an NRF24L01+ radio frequency module, and a Raspberry Pi for data reception and processing. This system can detect ambient noise levels, preprocess data using a Gaussian filter, and transmit information to an IoT platform for analysis and visualization. These systems showcase the versatility and scalability of IoT applications in urban settings. The Smart Parking Management System not only streamlines parking operations but also contributes to reducing traffic congestion and emissions by efficiently guiding drivers to available parking spaces. On the other hand, the E-nose air quality measurement system provides crucial data for environmental monitoring and enables authorities to take proactive measures in addressing air pollution issues. Overall, these IoT-based systems demonstrate the potential for technology to address critical issues in urban environments, such as parking management, air quality monitoring, and noise pollution. By leveraging IoT devices, wireless connectivity, and cloud-based platforms, these systems offer scalable and effective solutions for creating smarter and more sustainable cities.

ACKNOWLEDGEMENTS

The authors would like to express their sincere gratitude to Dr. Hitesh Mohapatra, Associate Professor at KIIT (Deemed to be) University, for his invaluable guidance and support throughout this work.

REFERENCES

[1] S. Talari, M. Shafie-khah, P. Siano, V. Loia, A. Tommasetti, and J. Catalão, "A Review of Smart Cities Based on the Internet of Things Concept", *Energies,* vol. 10, no. 4, p. 421, 2017. [http://dx.doi.org/10.3390/en10040421]

[2] V.I.B. Laya, and J. Markendahl, "Who is investing in machine-to-machine communications?", *Proc. 24th Eur. Reg. ITS Conf.,* 2013pp. 20-23

[3] "Performance evaluation of vehicle-based mobile sensor networks for traffic monitoring", *IEEE Trans. Vehicular Technol.,* vol. 58, no. 4, pp. 1647-1653, 2009. [http://dx.doi.org/10.1109/TVT.2008.2005775]

[4] S. J. Ramson, D. Bhavanam, S. Draksharam, R. Kumar, D. J. Moni, and A. A. Kirubaraj, "Radio Frequency Identification and Sensor Networks Based Bin Level Monitoring Systems—A Review", In: *Proc. 2018 4th Int. Conf. Devices, Circuits and Systems (ICDCS)* Coimbatore, India, 2018, pp. 16-17.

[5] S. Kaza, L. Yao, P. Bhada-Tata, and F. Van Woerden, *What a Waste 2.0: A Global Snapshot of Solid Waste Management to 2050.* World Bank Publications: Washington, DC, USA, 2018. [http://dx.doi.org/10.1596/978-1-4648-1329-0]

[6] H. Mohapatra, B. K. Mohanta, M. R. Nikoo, M. Daneshmand, and A. H. Gandomi, *MCDM-Based Routing for IoT-Enabled Smart Water Distribution Network.,* . [http://dx.doi.org/10.1109/JIOT.2022.3216402]

[7] C. Toma, A. Alexandru, M. Popa, and A. Zamfiroiu, "IoT Solution for Smart Cities' Pollution Monitoring and the Security Challenges", *Sensors (Basel),* vol. 19, no. 15, p. 3401, 2019. [http://dx.doi.org/10.3390/s19153401] [PMID: 31382512]

[8] T. Münzel, T. Gori, W. Babisch, and M. Basner, "Cardiovascular effects of environmental noise exposure", *Eur. Heart J.,* vol. 35, no. 13, pp. 829-836, 2014. [http://dx.doi.org/10.1093/eurheartj/ehu030] [PMID: 24616334]

[9] A.M. Said, A.E. Kamal, and H. Afifi, "An intelligent parking sharing system for green and smart cities based IoT", *Comput. Commun.,* vol. 172, pp. 10-18, 2021. [http://dx.doi.org/10.1016/j.comcom.2021.02.017]

[10] M. Humayun, M.S. Alsaqer, and N. Jhanjhi, "Energy Optimization for Smart Cities Using IoT", *Appl. Artif. Intell.,* vol. 36, no. 1, 2022. [http://dx.doi.org/10.1080/08839514.2022.2037255]

[11] W. Ejaz, M. Naeem, A. Shahid, A. Anpalagan, and M. Jo, "Efficient Energy Management for the Internet of Things in Smart Cities", *IEEE Commun. Mag.,* vol. 55, no. 1, pp. 84-91, 2017. [http://dx.doi.org/10.1109/MCOM.2017.1600218CM]

[12] C. Perera, A. Zaslavsky, P. Christen, and D. Georgakopoulos, "Sensing as a service model for smart cities supported by Internet of Things", *Trans. Emerg. Telecommun. Technol.,* vol. 25, no. 1, pp. 81-93, 2014. [http://dx.doi.org/10.1002/ett.2704]

[13] J. Jin, J. Gubbi, S. Marusic, and M. Palaniswami, "An Information Framework for Creating a Smart City Through Internet of Things", *IEEE Internet Things J.,* vol. 1, no. 2, pp. 112-121, 2014. [http://dx.doi.org/10.1109/JIOT.2013.2296516]

[14] S.N. Mohanty, J.M. Chatterjee, S. Satpathy, Ed., *Internet of Things and Its Applications.* Springer International Publishing, 2022. [http://dx.doi.org/10.1007/978-3-030-77528-5]

[15] I. Khajenasiri, A. Estebsari, M. Verhelst, and G. Gielen, "A Review on Internet of Things Solutions for Intelligent Energy Control in Buildings for Smart City Applications", *Energy Procedia,* vol. 111, pp. 770-779, 2017. [http://dx.doi.org/10.1016/j.egypro.2017.03.239]

[16] S. Lee, D. Yoon, and A. Ghosh, "Intelligent parking lot application using wireless sensor networks", *Proc. Int. Symp. Collab. Technol. Syst.,* vol. 19–23, pp. 48-57, 2008. [http://dx.doi.org/10.1109/CTS.2008.4543911]

[17] T. Nuortio, J. Kytöjoki, H. Niska, and O. Bräysy, "Improved route planning and scheduling of waste collection and transport", *Expert Syst. Appl.,* vol. 30, no. 2, pp. 223-232, 2006. [http://dx.doi.org/10.1016/j.eswa.2005.07.009]

[18] A.S. Syed, D. Sierra-Sosa, A. Kumar, and A. Elmaghraby, "IoT in Smart Cities: A Survey of Technologies, Practices and Challenges", *Smart Cities,* vol. 4, no. 2, pp. 429-475, 2021. [http://dx.doi.org/10.3390/smartcities4020024]

[19] B.L. Risteska Stojkoska, and K.V. Trivodaliev, "A review of Internet of Things for smart home: Challenges and solutions", *J. Clean. Prod.,* vol. 140, pp. 1454-1464, 2017. [http://dx.doi.org/10.1016/j.jclepro.2016.10.006]

[20] A. Saini, and A. Malik, "Routing in internet of things: A survey", *Commun. Comput. Syst.,* vol. I, pp. 855-860, 2017.

[21] H. Mohapatra, and A. K. Rath, *An IoT based efficient multi-objective real-time smart parking system., . International Journal of Sensor Networks*, vol. 37, no. 4, 2021. [http://dx.doi.org/10.1504/IJSNET.2021.119483]

[22] E. Casilari, R. Lora-Rivera, and F. García-Lagos, "A study on the application of convolutional neural networks to fall detection evaluated with multiple public datasets", *Sensors (Basel),* vol. 20, no. 5, p. 1466, 2020. [http://dx.doi.org/10.3390/s20051466] [PMID: 32155936]

[23] N. Mejri, "Reservation-based multi-objective smart parking approach for smart cities", in *Proc. 2016 IEEE International Smart Cities Conference (ISC2)*, Trento, Italy, 2016, pp. 1–6. [http://dx.doi.org/10.1109/ISC2.2016.7580840]

[24] H. Mohapatra, "Socio-technical Challenges in the Implementation of Smart City", [http://dx.doi.org/10.1109/3ICT53449.2021.9581905]

[25] Gongjun Yan, Weiming Yang, D.B. Rawat, and S. Olariu, "Smartparking: A secure and intelligent parking system", *IEEE Intell. Transp. Syst. Mag.,* vol. 3, no. 1, pp. 18-30, 2011. [http://dx.doi.org/10.1109/MITS.2011.940473]

[26] P. Melnyk, S. Djahel, and F. Nait-Abdesselam, "Towards a Smart Parking Management System for

Smart Cities",
[http://dx.doi.org/10.1109/ISC246665.2019.9071740]

[27] M. Taştan, and H. Gökozan, "Real-time monitoring of indoor air quality with internet of things-based e-nose", *Appl. Sci. (Basel),* vol. 9, no. 16, p. 3435, 2019.
[http://dx.doi.org/10.3390/app9163435]

[28] R. Vázquez-López, E.V. Vázquez-Carmona, J. Carlos Herrera-Lozada, K. Miranda, and J. Sandoval-Gutierrez, "A proposed embedded system for indoor noise pollution monitoring via IoT", *IEEE ANDESCON, Barranquilla, Colombia,* vol. 2022, pp. 1-6, 2022.
[http://dx.doi.org/10.1109/ANDESCON56260.2022.9989829]

[29] A. Rauf, A. Guru, and H. Mohapatra, "Exploring the Social Model of Smart Home Devices", *Adv. Hum. Soc. Asp. Technol.,* no. Sep, pp. 233-256, 2024.
[http://dx.doi.org/10.4018/979-8-3693-9235-5.ch012]

[30] S. Mukherjee, and H. Mohapatra, "Performance Analysis of Different MANET Routing Protocols",
[http://dx.doi.org/10.1109/ICITIIT61487.2024.10580492]

CHAPTER 9

Next-Generation Urban Mobility Management: A Framework for Intelligent Traffic Control

Sujithra Thandapani[1,*], Durai Selvaraj[1], Mohamed Iqbal M.[3], D. Jaganathan[4] and V.C. Bharathi[5]

[1] *Department of CSE, Manipal Institute of Technology, Manipal Academy of Higher Education, Manipal, Karnataka, India*

[2] *Department of CSE, Vel Tech Rangarajan Dr. Sagunthala R&D Institute of Science and Technology, Avadi, Chennai, India*

[3] *Department of Information Technology, Sri Sivasubramaniya Nadar College of Engineering, Chennai, 603110, India*

[4] *Department of Artificial Intelligence, Madanapalle Institute of Technology & Science, Madanapalle, India*

[5] *School of Computer Science & Engineering, VIT-AP University, Amaravati, Andhra Pradesh, India*

Abstract: Traffic congestion is exacerbated by the proliferation of vehicles, ineffective traffic management practices, and inadequate road infrastructure. Especially in cities, people are struggling to have quick mobility. Improper traffic management leads to longer transit times and contributes to fuel waste, air pollution, and increased accident risk. Conventional traffic control systems set the service time for each lane based on its nature. For example, fast lanes may get more service time than slow ones. It is a static system. It uses fixed timing to service all the lanes irrespective of traffic density. It results in increased traffic in a specific lane. In most countries, automated systems are unavailable to service emergency vehicles (ambulances, firefighting vehicles, mobile medical units, blood donation vehicles, and VIP vehicles) due to the enormous implementation and maintenance costs. In this chapter, a low-cost edge device is proposed for an intelligent traffic control system to address this issue. The proposed system focuses on dynamic signaling and automatic green enablers for emergency vehicles. Dynamic signaling controls the timing of the traffic control system based on the lane's traffic density, which is measured by the You Only Look Once (YOLO V3) algorithm. An automatic green light enabler enables the green light in advance for the lane where the emergency vehicle's presence is 750 meters away from the traffic signal. It deals with various types of emergency vehicles, such as ambulances, firefighting vehicles, and mobile medical units. Computer vision and communication technologies are utilized in designing a cost-effective and innovative traffic control system. It assists

* **Corresponding author Sujithra Thandapani:** Department of CSE, Manipal Institute of Technology, Manipal Academy of Higher Education, Manipal, Karnataka, India; E-mail: sujimecse@gmail.com

Suneeta Satpathy, Sachi Nandan Mohanty & Subhendu Kumar Pani (Eds.)

in efficiently managing traffic. It also services emergency vehicles at road intersections. Decreasing the average waiting time for emergency vehicles at road intersections contributes to society by saving lives.

Keywords: Computer vision, Emergency vehicles, Edge devices, IoT, LORA, Traffic.

INTRODUCTION

Due to the rapid evolution of the social economy and the advancement of science and technology, urban areas have undergone significant expansion. Consequently, the issue of traffic congestion has become exceedingly severe. In conventional traffic control systems, traffic lights and signs are limited to providing static and restricted guidance, making it challenging to optimize road resources [1]. We could address the problem by upgrading and optimizing the urban traffic system with the help of effective coordination and cooperation. To scientifically alleviate these pressures and improve convenience for urban residents, it is imperative to swiftly integrate new technologies to transform and enhance traditional traffic modes. Smart city traffic systems represent a significant breakthrough in this regard. Especially with the integration of Internet of Things (IoT) technology, advancements in new smart city traffic systems show significant potential for easing urban traffic congestion. Smart city transportation systems in urban areas address several practical issues by providing information about road conditions, traffic details, alternative routes, and service guidance, among others [2].

Leveraging Internet of Things (IoT) technology, the smart city traffic system adeptly gathers extensive road and vehicle data within the urban landscape. Subsequently, it dynamically calculates optimal route planning to facilitate the smooth flow of urban traffic [3]. Furthermore, integrating IoT technology is pivotal for augmenting the effectiveness of traffic command and scheduling. IoT-enabled traffic command and dispatch can more effectively anticipate and mitigate widespread traffic congestion, consequently ensuring a smoother travel experience for residents [4].

Numerous researchers are currently focusing on advancing the development of smart city transportation infrastructure. Research in this field is still in its early stages, yet it employs cutting-edge technology [5]. By implementing intelligent control systems, urban traffic signal lights can be managed intelligently, reducing traffic congestion [6]. Some notable works in traffic control systems are discussed below. For example, a study [7] reported that students at India's Chandigarh University have developed an intelligent, sensor-based traffic control system designed to give priority passage to ambulances and fire trucks. The presence of emergency service vehicles on the road is identified by RF (Radio Frequency) transmitters and receivers installed in the cars. The sound detection system in ambulances and fire trucks sends signals to the traffic controller, changing the signal to green while keeping the other signals red [8]. The RF receiver can detect the sound frequency of an emergency vehicle within a range of 328-656 feet (100-200m). Chitkara University students, in collaboration with the Mohali traffic police, have developed a 3D innovative traffic signal system aimed at regulating traffic signals using a wireless sensor system and providing a smart bird's-eye view. The system dynamically adjusts traffic movement based on the volume of approaching traffic through a self-sensing technique. It prioritizes green passages for ambulances and fire tenders while managing traffic flow. This innovative system has been deployed at the traffic crossing near Quark City on Airport Road, with an estimated cost of Rs 50,000.

Bengaluru, India, in partnership with the Japan International Cooperation Agency (JICA), successfully implemented an advanced traffic information and management system, involving an investment of Rs 72.86 crore [9]. The system has been deployed at 29 junctions, including prominent locations, such as G Road, Old Madras Road, and Hosur Road. It operates by automatically adjusting signals based on the volume of vehicles at each junction. The system continuously analyses real-time traffic congestion by being equipped with CCTV cameras to gauge vehicle length and size, and sensors positioned at intervals of 50, 100, and 150 meters along roads. It updates the integrated traffic management center accordingly. According to *Traffic Management to Get 'Smarter' in City* (2013) [10], the Hyderabad Traffic Integrated Management System (H-TRIMS) in India implemented intelligent traffic signals at the Jubilee Hills check post in Hyderabad. These signals regulate the duration of red and green lights at traffic junctions based on real-time vehicular traffic density. Cameras installed at this junction continuously monitor traffic density and automatically adjust the duration of green and red lights accordingly. The state government has allocated Rs. 66.50 crore for the H-TRIMS project.

As reported by Shivakumar, in 2019 [11], an intelligent traffic control system was deployed at 12 locations in Chennai city, India, to manage traffic flow efficiently. In partnership with the Japan International Cooperation Agency (JICA), this project has secured funding of Rs 660 crore for the implementation of an Intelligent Transport System. The system utilizes CCTV-based traffic management and incorporates bright signals, which predict traffic volume from upstream intersections to alleviate congestion. These smart signals automatically adjust timings based on traffic volume, effectively reducing congestion. Dubai launched a state-of-the-art, innovative traffic control system powered by artificial intelligence (AI) designed to predict and manage vehicle flow, formulate rapid response plans, and make real-time decisions in the event of accidents [12]. This system includes the installation of 116 cameras, 100 incident detection systems, 114 transit time and speed computing devices, and 17 Road Weather Information Systems (RWIS) [13]. These components collectively reduce accident detection time and forecast traffic congestion. Launched by the Government of Dubai, this initiative represents one of the largest and most advanced traffic control centers globally. Singapore has developed a next-generation innovative traffic light control system utilizing AI and sensors, initially implemented at Corporation Road and Boon Lay Way [14]. This system is designed to detect the physical presence of vehicles and pedestrians. By leveraging real-time datasets from advanced technologies, such as the Global Navigation Satellite System (GNSS) and autonomous vehicles, AI and predictive capabilities can aid in developing intelligent traffic light algorithms [15]. Paris has developed fully adaptive traffic algorithms that enable adjustments to the duration of green traffic lights based on traffic density [16]. This innovative system has been installed at several Val-de-Marne region intersections. However, a significant number of sensors must be installed at intersections, averaging between 20,000 and 25,000 sensors for every 2,000 traffic light controllers, to predict traffic congestion accurately. China has developed an innovative city traffic management system utilizing big data-driven AI-based technologies. This system offers various transportation infrastructure enhancements, such as traffic flow measurements, smart traffic signals, and traffic management programs for maintenance scheduling and system assessments. Notably, over 1,200 intersections nationwide have implemented smart traffic signals as part of this initiative [17]. The city of Pittsburgh, Pennsylvania (USA), has implemented innovative traffic technology leveraging Artificial Intelligence. This system utilizes radar sensors and cameras to recognize traffic activity and optimize traffic flows at 50 intersections throughout Pittsburgh.

From a national and international standpoint, it has been observed that the existing innovative traffic control system utilizes GPS (Global Positioning System) to track vehicles. With the help of GPS, smart sensors, CCTV cameras, and cloud infrastructure, the traffic density and control system are managed and control. However, it results in increased implementation and maintenance costs. Due to the high implementation cost, governments are unable to implement automated systems in all areas. However, it can be addressed by designing low-cost devices.

In many countries with high traffic volumes, traffic control primarily depends on fixed-time signals. Conversely, major cities in certain developed nations manage traffic through centrally controlled systems. Consequently, it can be concluded that current traffic management systems continue to operate in a centralized manner.

Shortcomings in the Existing System

The intelligent traffic control system needs an assertive cloud and good internet services. It requires a higher maintenance cost than the implementation cost. The survey report shows that no cloud-free, cost-effective automatic traffic control system exists for managing traffic density and servicing emergency vehicles. In Hyderabad and Bangalore, India, a smart traffic control system is installed only for servicing ambulances, not for all types of emergency vehicles.

PROPOSED SYSTEM

To address the shortcomings in the existing system, we proposed a low-cost edge device for an intelligent traffic control system. It is a device that works based on edge computing technology. It primarily focuses on enabling and scheduling automatic green lights, as well as controlling traffic systems based on real-time traffic density. In general, conventional systems are centralized. A server/ cloud is needed to do the analysis, which helps to control and manage the traffic system. If the internet or a centralized system fails, the entire infrastructure will be affected. It also incurs more maintenance costs. To address this problem, the proposed system makes decisions on the device itself rather than relying on the server or cloud, thereby reducing latency and computational overhead.

Proposed Edge device

Fig. (**1**) shows the proposed edge device installed at a traffic signal located at a road intersection. It comprises a LoRa communication module for sending and receiving signals between emergency vehicles and the traffic control system, as well as cameras for capturing road traffic conditions. YOLO V3 takes real-time camera images from all directions as input for computing traffic density. With the help of a communication module, we can also connect edge devices with the server for historical analysis/future forecasting if needed. The proposed system works on two primary components, automatic green light enabler and dynamic signaling, which are discussed as follows:

Automatic Green Light Enabler

The primary objective of the automatic green light enabler is to activate the green light in advance for the lane where an emergency vehicle is detected 750 meters away from the traffic signal. To service different types of emergency vehicles, a dedicated frequency is assigned to every emergency vehicle for unique identification. Emergency vehicles are classified as homogeneous (all the same type of vehicles, such as our ambulances) and heterogeneous (different types of emergency vehicles, including ambulances, fire tenders, mobile medical units, *etc.*). In general, once an intelligent traffic control system identifies an emergency vehicle using its frequency, it immediately turns the green light on for the specific lane.

Fig. (1). Flow regimes and governing equations.

In real-time, we have several challenges in servicing emergency vehicles, which are listed below.

Case 1: Fig. (**2**) shows that the lane may have a single emergency vehicle, and a green light is enabled immediately for that lane.

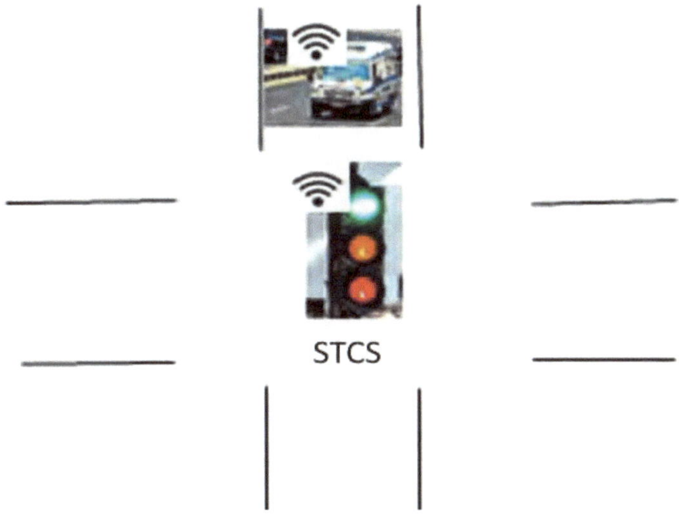

Fig. (2). Single emergency vehicle. lane

Case 2: Fig. (**3**) shows that more than one ambulance may arrive in different lanes. The distance of all ambulances from the traffic signals is determined using the received signal strength.

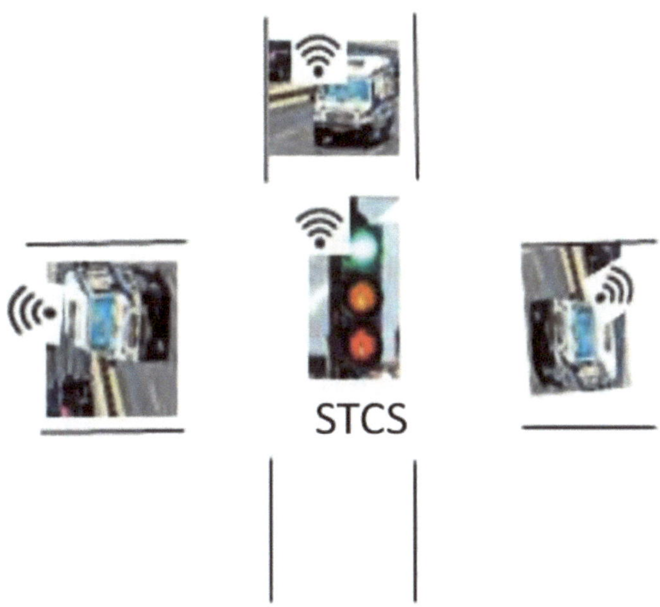

Fig. (3). Multiple emergency vehicles lane.

Case a: The distance of the ambulance from the traffic signal is different. The ambulance closer to the signal will get the highest priority compared to the others.

Case b: All ambulances are at equal distances. The ambulance that arrives first will be serviced first.

Case 3: Fig. (**4**) shows that heterogeneous emergency vehicles may arrive at different lanes. Based on the priority, emergency vehicles will be serviced. According to global traffic rules, ambulances will receive the highest priority over fire tenders. Fire tenders will be given the highest priority, not VIPs or police vehicles.

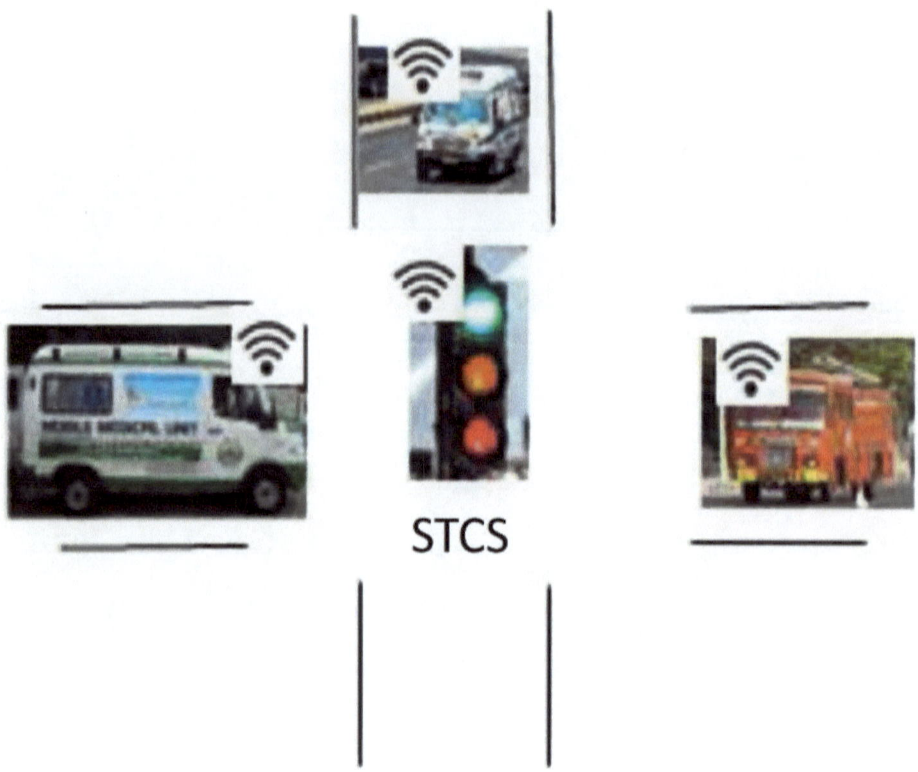

Fig. (4). Heterogeneous emergency vehicles lane.

Dynamic Signaling

Dynamic signaling is applicable only when emergency vehicles are not available in all lanes. If an emergency vehicle is present in any lane, priority is given to emergency vehicles. In the dynamic scheduling algorithm shown in Fig. (**5**), the timing of the traffic control system is adjusted automatically based on the live traffic density. Images are taken periodically from all the signal lanes to measure traffic density. YOLO V3 is utilized to measure traffic density. It utilizes a CNN to detect objects within the bounding boxes. It takes images from all directions as input. After that, scheduling is done to adjust the traffic signal's timing based on the measured traffic density. Based on traffic density, lanes are classified into fast lanes, medium segment lanes, and slow lanes. We set different threshold values for each lane based on the object count. If the object count is higher, they will receive more green time than the lane with a lower object count, as discussed in the dynamic scheduling algorithm. All lanes are serviced in a circular manner; only the green light timings are adjusted based on traffic density.

Proposed Dynamic Scheduling Algorithm

Input: Results of YOLO v3 (Objects present in every lane) are given as input.

Variables: t_1,t_2,and t_3 are thresholds set it for every lane. Threshold is minimum number of objects.

t_m- Measured traffic density (Number of objects are present in a particular lane)

l_f- fast lane

l_m- medium segment lane

l_s- slow lane

dg- duration of the green signal

δt_1, δt_2, δt_3 are time assigned for green signalling.

Constraints:

$\delta t_1 > \delta t_2 > \delta t_3$, $t_1 > t_2 > t_3$

Logic for classifying the lane and green signal setting.

```
if (tm > t1)
    lane= lf
        dg = δt1
if (t1 < tm > t2)
    lane= lm
        dg = δt2
if (t2 < tm > t3)
    lane= ls
        dg = δt3
```

Fig. (5). Dynamic scheduling algorithm.

Based on the traffic pattern analysis, we assigned different service times to various lanes. All the lanes are serviced with varying service times based on captured traffic density. A lane with more vehicles is serviced first, then the next congested lane, and so on.

IMPLEMENTATION AND EXPERIMENTAL RESULTS

The proposed edge device is integrated with the existing traffic signal system. The edge device comprises Arduino UNO for processing, a camera module for capturing live traffic images at road intersections, and LORA for sending and receiving signals to and from the traffic control system and emergency vehicles. In this study, we designed an automatic green light enabler for servicing emergency vehicles and an innovative system for dynamic signaling based on traffic density.

Automatic Green Light Enabler

Vehicles are classified as emergency and normal vehicles for the design of an automatic green light enabler. Emergency vehicles are classified into high-priority, medium-priority, low-priority, and priority-less. In the developed system, ambulances are assigned high priority, fire tenders are assigned medium priority, mobile medical units, blood donation vehicles, and VIP vehicles are assigned low priority, and normal vehicles are treated as priority-less vehicles. We assigned a unique frequency to every emergency vehicle to distinguish between them and establish communication between the edge device and the emergency vehicle, as shown in Figs. (**6**, **7**, and **8**). The ISM band is used for assigning frequencies. We fixed four LORA modules in all directions of the road intersection. It is simple and cost-effective. It does not require any internet services and results in reduced implementation and maintenance costs. Hence, we can implement this system in many places. With the help of this communication module, we could find the lane with an emergency vehicle. The maximum range of the communication module is 1km. With the help of this device, we could notify the edge device integrated with traffic signals regarding the availability of emergency vehicles. As discussed previously, it helps to turn the green light in advance for the lane where the emergency vehicle's presence is 750 meters away from the traffic signal. If more than one lane has emergency vehicles, scheduling is based on the communication range and the type of emergency vehicle. The emergency vehicle closer to the signal will be serviced first, then the next one, and so on. If we have emergency vehicles in more than one lane, and all the emergency vehicles are within the same communication range, they will be serviced in a circular manner. It results in reduced average waiting time. We compared the results of the proposed system with those of the existing conventional traffic control system for validation purposes. The experimentation results revealed that the proposed system yielded better results than the traditional traffic control system, as presented in Table 1.

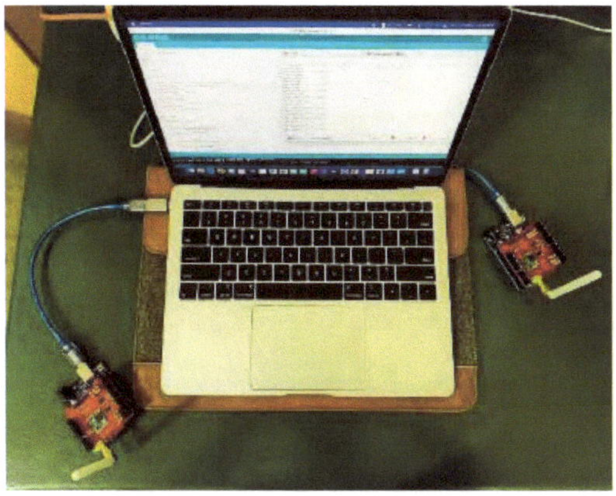

Fig. (6). Testbed for detecting and classifying emergency vehicles.

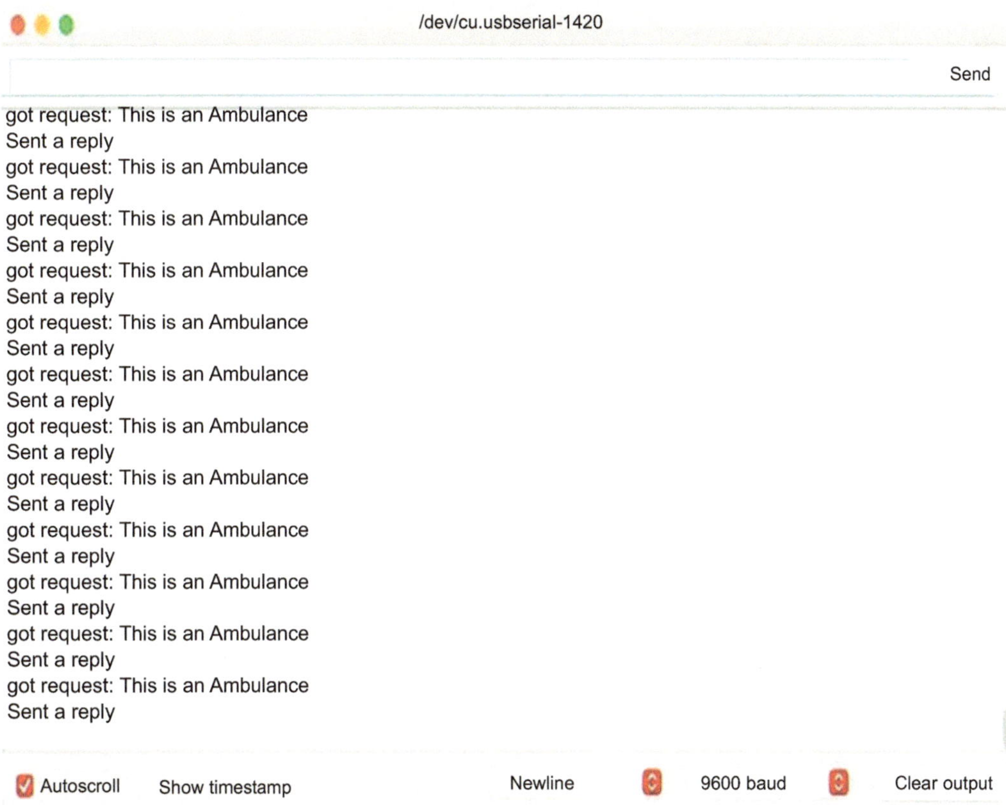

Fig. (7). Emergency vehicle classification results.

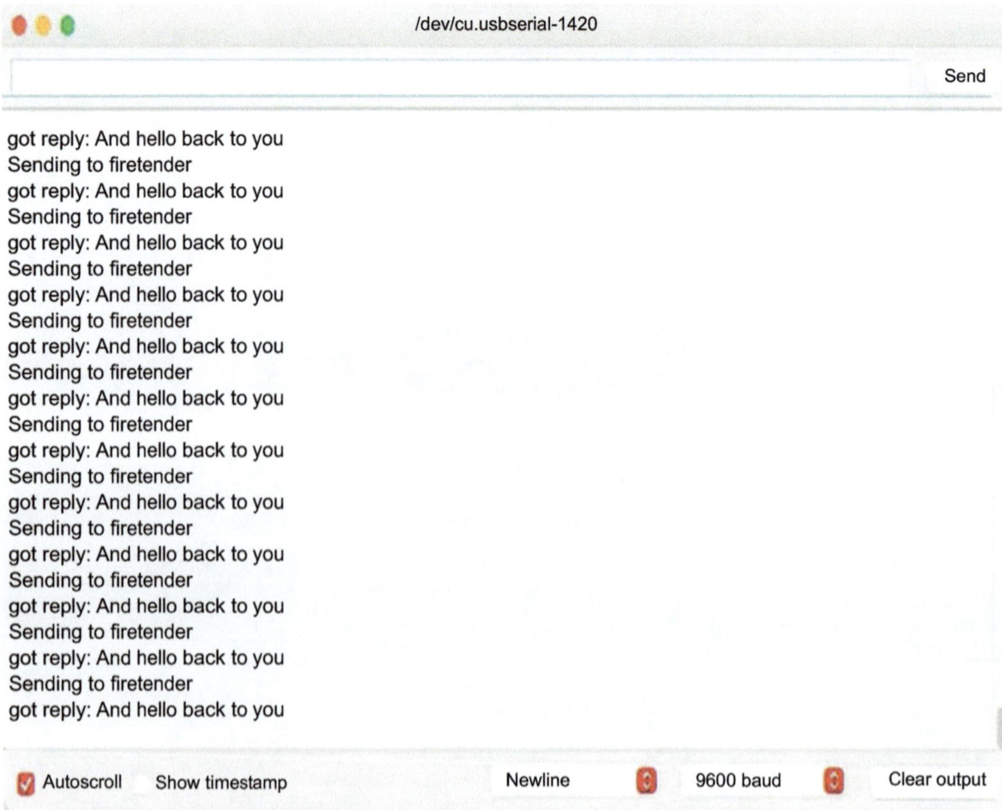

Fig. (8). Emergency vehicle classification results.

Table 1. Conventional system *Vs.* proposed system.

Algorithm for Green Light Enabling	Time (Cycles)	Delay (Cycles)
Conventional system using round-robin	4000	26.2
	8000	22.3
	12000	20.8
Proposed System	4000	10.4
	8000	6.3
	12000	3.2

Dynamic Scheduling

An innovative system is also designed for regulating the traffic flow based on live traffic density. For dynamic scheduling, computer vision techniques are employed using a camera module to capture real-time traffic information. Simple cameras are used for data collection instead of HD cameras. It captures only images of the lane traffic instead of live streaming (videos), resulting in reduced computational complexity. As previously discussed, YOLO V3 is used to calculate traffic density, which utilizes a CNN to detect objects within bounding boxes and then takes images from all directions as input. The dataset is collected from Kaggle. It contains 2,310 images of cars, two-wheelers, autorickshaws, buses, and lorries. Moreover, 80% of the dataset is used for training, and 20% is used for testing the model. The model yields 93.5% classification accuracy. It successfully classifies different types of vehicles, as shown in Figs. (**9**, **10**, and **9**).

Fig. (9). YOLO results for 2-wheeler.

Fig. (10). YOLO results for 4-wheeler.

Fig. (11). YOLO results for 2- and 4-wheelers.

Based on the traffic density, traffic signal timings are adjusted. All the lanes are serviced circularly. However, the timing of the green signal is adjusted based on the traffic density computed by YOLO V3. The high-density lane is scheduled with more green signaling time than the low-density lane. Edge impulse is utilized as an interactive environment to transform the edge device into an intelligent system. With the help of edge impulse, the model is trained for computing traffic density. Then, the edge device is connected to the edge impulse for real-time analysis (Table **2**). The YOLO V3 results are taken at time instant t1 for the different objects.

Table 2. YOLO V3 results taken at time instant t1.

Type of Vehicles	Object Count
Car	10
Bus	3
Autorickshaw	1
Two-wheeler	17

CONCLUSION

A smart traffic control system enables emergency vehicles to reach their destinations quickly. The proposed system reduces the time spent on traveling. It is particularly helpful for urban dwellers who spend a significant amount of time traveling. As the proposed system reduces the average waiting time of vehicles at road intersections, it can reduce fuel wastage. It also helps to increase road safety. Pollution prevention and carbon footprint are hot topics now. Optimizing the traffic control system's operation will lead to a reduction in the carbon footprint.

As low-cost devices are designed for the system, we can deploy them in many places. It is helpful for the traffic department in handling and managing traffic. Additionally, it will be valid for the medical community in saving people's lives.

REFERENCES

[1] L. Huang, X. Yuan, J. Zhang, N. Zhang, J. Li, and L. Wang, "Research on internet of things technology and its application in building smart communities", *J. Phys. Conf. Ser.*, vol. 1550, no. 2, p. 022029, 2020.
[http://dx.doi.org/10.1088/1742-6596/1550/2/022029]

[2] N.A. Khan, N.Z. Jhanjhi, S.N. Brohi, R.S.A. Usmani, and A. Nayyar, "Smart traffic monitoring system using Unmanned Aerial Vehicles (UAVs)", *Comput. Commun.*, vol. 157, pp. 434-443, 2020.
[http://dx.doi.org/10.1016/j.comcom.2020.04.049]

[3] F. Outay, H.A. Mengash, and M. Adnan, "Applications of unmanned aerial vehicle (UAV) in road safety, traffic and highway infrastructure management: Recent advances and challenges", *Transp. Res. Part A Policy Pract.*, vol. 141, pp. 116-129, 2020.
[http://dx.doi.org/10.1016/j.tra.2020.09.018] [PMID: 33024357]

[4] M.M. Rathore, A. Ahmad, A. Paul, and S. Rho, "Urban planning and building smart cities based on the Internet of Things using Big Data analytics", *Comput. Netw.*, vol. 101, pp. 63-80, 2016.
[http://dx.doi.org/10.1016/j.comnet.2015.12.023]

[5] A. Chowdhury, "Priority based and secured traffic management system for emergency vehicles using IoT".
[http://dx.doi.org/10.1109/ICEMIS.2016.7745309]

[6] M.A.A. Mamun, M.A. Hannan, A. Hussain, and H. Basri, "Theoretical model and implementation of a real time intelligent bin status monitoring system using rule based decision algorithms", *Expert Syst. Appl.*, vol. 48, pp. 76-88, 2016.
[http://dx.doi.org/10.1016/j.eswa.2015.11.025]

[7] A. Frost, *Self-powered smart traffic control system developed by Indian students.*, 2020.https://www.traffictechnologytoday.com/news/intersections/self-powerin--smart-traffic-control-system-developed-by-indian-students.html

[8] https://government.economictimes.indiatimes.com/news/digital-india/digital-india-mohali-gets-i-dias-first-3-d-smart-traffic-signal-system/70510522 n.d.

[9] C. Belagere, https://www.newindianexpress.com/cities/bengaluru/2018/sep/15/bengaluru-japanes--tech-to-help-in-traffic-management-1872213.html

[10] https://www.thehindu.com/news/cities/Hyderabad//article60032920.ece

[11] C. Shivakumar, *Chennai may get smart traffic signals in 12 junctions.*, 2019.https://www.newindianexpress.com/cities/chennai/2019/aug/07/chennai-may-get-smart-traffic-signals-in-12-junctions-2015560.html

[12] *Sheikh Mohammed opens smart traffic monitoring system.*, 2020.https://theuaenews.com/2020/11/10/sheikh-mohammed-opens-smart-traffic-monitoring-system/

[13] *Cruise-ing along the roads of Singapore with AI.*.https://www.a-star.edu.sg/i2r/news/cruise-ing-al-ng-the-roads-of-singapore-with-ai n.d.

[14] https://www.spie.com/en/news/smart-urban-mobility-spie-integrates-traffic-prediction-val-de-marne-region n.d.

[15] *DiDi introduces smart transportation solution for traffic management.*, 2018.https://www.businesswire.com/news/home/20180125006404/en/DiDi-Introduces-Smart%20Transportation-Solution-for-Traffic-Management

[16] https://smartcityhub.com/mobility/smart-traffic-control/

[17] M. Jonnalagadda, S. Taduri, and R. Reddy, "RealTime traffic management system using object detection based signal logic", *2020 IEEE Applied Imagery Pattern Recognition Workshop (AIPR),* 2020pp. 1-5 [http://dx.doi.org/10.1109/AIPR50011.2020.9425070]

CHAPTER 10

Predicting Customer Churn in the Telecom Industry with Machine Learning Techniques for Improved Retention Strategies

Smital Anandrao Patil[1], **Rushi Patel**[2], **Swati Shukla**[3,*], **Yeswanth Satya Prasad**[4] and **Yagnesh Challagundla**[5]

[1] *School of Computer Science, KIT's College of Engineering, Kolhapur-416234, Maharashtra, India*

[2] *Department of Electronic and Telecommunication Engineering, Indian Institute of Technology - IIT, Kharagpur-721302, West Bengal, India*

[3] *School of Electronics Engineering (SENSE), VIT-AP University, Amaravati- 522237, Andhra Pradesh, India*

[4] *School of Computer Science and Engineering (SCOPE), VIT University, Vellore- 632014, Tamil Nadu, India*

[5] *School of Computer Science and Engineering (SCOPE), VIT-AP University, Amaravati, 522237, Andhra Pradesh, India*

Abstract: In the commercial world, customers are paramount, and understanding their behavior patterns can lead to highly impactful business decisions. Customer churn, defined as the rate at which consumers switch from one business to another, presents a significant obstacle, especially for businesses that rely on subscription models, such as telecommunications companies. Accurately forecasting customer turnover enables firms to effectively implement targeted retention initiatives and minimize income loss. The objective of this study is to employ logistic regression in order to forecast customer attrition for a telecoms company. We utilized a dataset that comprised comprehensive client information, encompassing demographics, service usage, and payment records. Through a thorough exploratory data analysis (EDA), we identified important patterns and trends that are directly linked to customer attrition. We performed data preprocessing by addressing missing values, encoding category variables, and normalizing numerical features. A robust statistical technique, specifically designed for binary classification, was employed to construct a predictive model. The discovered key predictive variables included tenure, monthly charges, and contract type. Our data indicate that clients who have been with us for a shorter period have higher monthly rates, and those on month-to-month contracts are more likely to discontinue their services. The regression model attains the highest accuracy, surpassing other machine learning techniques in terms of overall predictive capability.

* **Corresponding authors Swati Shukla:** School of Electronics Engineering (SENSE), VIT-AP University, Amaravati- 522237, Andhra Pradesh, India; E-mail: swati.shukla@vitap.ac.in

Suneeta Satpathy, Sachi Nandan Mohanty & Subhendu Kumar Pani (Eds.)

This study highlights the efficacy of this instrument as a valuable tool for forecasting customer turnover, enabling firms to enhance customer satisfaction and reduce churn rates by making informed decisions based on data.

Keywords: Classification models, Customer churn, Customer retention, Data analysis, Machine learning, Predictive modeling, Telecom industry.

INTRODUCTION

Customer churn, also known as attrition, is the act of losing clients or customers and is a vital measurement for businesses that rely on subscriptions, such as telecommunications companies. Anticipating customer attrition can have a substantial influence on a company's capacity to maintain its client base. The objective of this study is to forecast customer attrition for a telecommunications business by employing logistic regression. The data utilized in this investigation was obtained from a Kaggle dataset, consisting of 7043 distinct client entries with 21 characteristics, encompassing demographic details, services utilized, and monthly costs. Precise churn prediction enables telecom businesses to formulate focused strategies for retaining customers who are likely to leave, resulting in decreased turnover rates and enhanced profitability. This project encompassed various crucial stages: data acquisition, preprocessing, conversion, and modeling. Initially, the data underwent a process of cleaning and recoding to rectify any discrepancies and missing data. Descriptive statistics and visualizations provided a comprehensive analysis of the data's distribution, revealing any potential relationships between variables. After conducting the exploratory data analysis, we used logistic regression to create a model for predicting the binary outcome of churn. Logistic regression was used due to its straightforwardness and efficacy in managing categorical outcome variables. This approach enabled us to measure the magnitude of the influence of different predictor variables, such as tenure, contract type, and total costs, on the probability of customer turnover. The performance of the logistic regression model was assessed using common measures, including accuracy, precision, recall, and the confusion matrix.

The primary objective of this research is to identify the key indicators of customer churn and develop a reliable model that accurately predicts churn occurrences. Telecom firms may effectively manage customer churn by comprehending the reasons that lead to it. This understanding enables them to take proactive measures to address issues and create retention tactics, thereby ensuring a stable client base. This study demonstrates the practical application of logistic regression in predictive modeling and highlights the importance of making informed decisions based on data in the telecommunications industry.

RELATED WORK

Smith *et al.* [1] proposed a machine learning approach for forecasting customer turnover in the telecommunications industry. Their research explored various algorithms and emphasized the importance of selecting the optimal one based on specific data attributes. Lee *et al.* [2] performed a comparative analysis of machine learning techniques for predicting customer attrition in telecommunications networks. Their study evaluated the efficacy of several algorithms and provided valuable insights into their advantages and limitations for this specific application. Garcia [3] highlighted the significance of machine learning in enhancing client retention within the telecoms sector. This study examined the overarching benefits of employing machine learning to predict customer churn and refine strategies for customer retention. Chen and Zhang [4] concentrated on enhancing customer lifetime value in the telecommunications industry by utilizing machine learning to forecast churn. Their research examined the application of attrition prediction algorithms to enhance customer value for telecommunications firms. Khan *et al.* [5] presented a thorough examination of how artificial intelligence (AI) is used in managing customer turnover in the telecom industry. This paper examined various artificial intelligence (AI) methods used in churn prediction and their impact on client retention efforts. Another study [6] investigated the application of machine learning techniques to identify consumers at a higher risk of churn in telecom firms. At a conference, Patel (2023) [7] presented a discussion on the use of data-driven techniques for customer retention and the application of machine learning to anticipate churn in the telecommunications industry. Miller *et al.* [8] emphasized the cost-saving potential of machine learning in predicting customer turnover in the telecommunications industry, particularly highlighting its financial advantages for telecom corporations. Brown (2022) [9] explored the application of machine learning in the telecom industry for the purpose of customer retention. The author presented it as a means to anticipate and forecast future customer behavior. Williams (2020) [10] highlighted the significance of machine learning in ensuring customer satisfaction in the telecom industry by accurately anticipating customer turnover and proactively resolving any future problems. In their study, Gupta and Singh (2022) [11] investigated the application of deep learning techniques to predict customer churn in the telecommunications industry, highlighting the significant capabilities of deep learning in this specific field. Yang *et al.* (2022) [12] presented a cost-sensitive machine learning method for predicting customer attrition in telecommunications services. This study examined the financial implications of customer attrition and sought to refine prediction models accordingly. Manzoor *et al.* (2024) [13] presented an extensive examination of machine learning techniques for predicting customer attrition, providing suggestions for business professionals.

Szelag and Slowinski (2024) [14] primarily aimed to elucidate and forecast customer attrition by employing transparent criteria generated from customer data, thereby providing valuable insights into the underlying causes of churn. Gurung *et al.* (2024) [15] investigated artificial intelligence (AI)-driven churn prediction models that could be used in corporate marketplaces, including the telecom sector. Liu *et al.* (2024) [16] proposed a novel approach employing extreme gradient boosting and Bayesian optimization for predicting churn in a profit-driven manner, with the primary goal of maximizing financial gains. De Caigny *et al.* (2024) [17] suggested a hybrid black-box classification method for predicting customer turnover. Their technique incorporated interpretable analysis to achieve a balance between model accuracy and interpretability. Jiang *et al.* (2024) [18] proposed a profit-driven weighted classifier that enhanced interpretability in customer churn prediction, building upon the work of Liu *et al.* (2024) but with a specific emphasis on interpretability.

METHODOLOGY

The process of predicting customer churn in a telecom company using regression involves multiple essential steps, each contributing to the thorough research and development of the model. For better understanding, we have constructed a PlantUML tree (Fig. **1**) below, which depicts the exact flow of the models used.

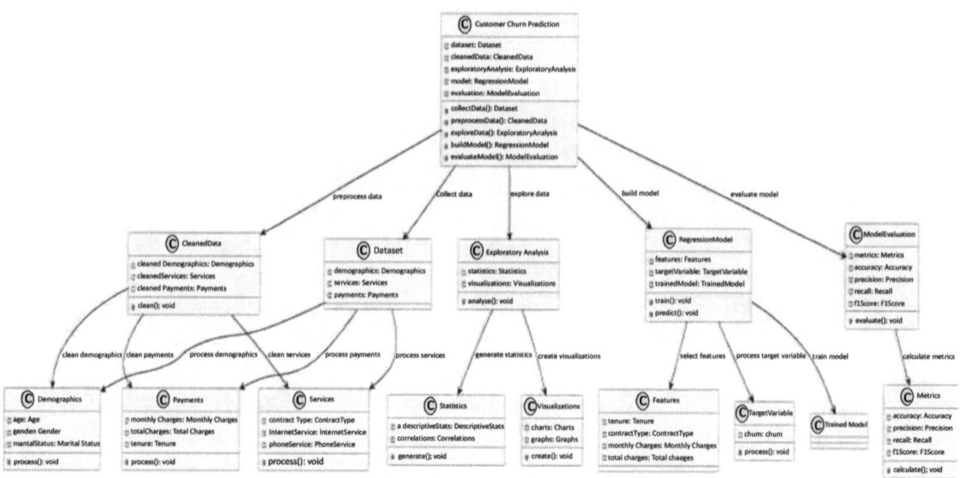

Fig. (1). Methodology flowchart: a visual guide to our process.

Data Collection

The dataset used for this analysis consists of 7,043 entries, representing distinct consumers. It has 21 columns, including the target variable 'Churn'. The features encompass demographic data, utilized services, and monthly fees.

Data Preprocessing

Initially, the data was loaded and purified to ensure its quality and uniformity. This entailed managing null values, transforming categorical variables, and removing extraneous columns, such as customer IDs. The variables SeniorCitizen, MultipleLines, and several internet-related services were restructured to streamline the analysis. In addition, the data were modified to ensure they could be used with the logistic regression model. This involved transforming category variables into dummy variables and scaling numerical features.

Exploratory Data Analysis (EDA)

A preliminary data analysis was performed to comprehend the patterns and trends present in the dataset. Descriptive statistics and visualizations were created to emphasize important features of the data. For example, bar plots were utilized to analyze demographic data, histograms were employed to examine numerical aspects, and correlation matrices were utilized to detect correlations between variables. The study revealed notable findings, which included a robust association between TotalCharges and tenure, as well as between TotalCharges and MonthlyCharges.

Development of a Model

The regression models were selected for their efficacy in solving binary classification problems. The dataset was partitioned into training and test sets.

Evaluation of the Model

The model's performance was evaluated using a confusion matrix, which offered insights into the number of true positives, true negatives, false positives, and false negatives. The model's predictive power was evaluated by calculating the accuracy, precision, recall, and F1-score. The model achieved the highest accuracy, surpassing alternative machine learning techniques, such as decision trees and random forests.

Importance of Features

The model was used to identify significant predictors of churn based on their p-values. The primary variables considered were the length of tenure, the contract status, and total costs. Month-to-month contracts and higher total charges were found to be significant predictors of customer churn.

By conducting thorough data pretreatment, exploratory data analysis (EDA), and regression modeling, we successfully created a robust model that accurately predicts customer attrition. This approach not only yielded a strong prediction model but also provided useful insights into the factors that influence customer attrition in the telecommunications business (Fig. **2**).

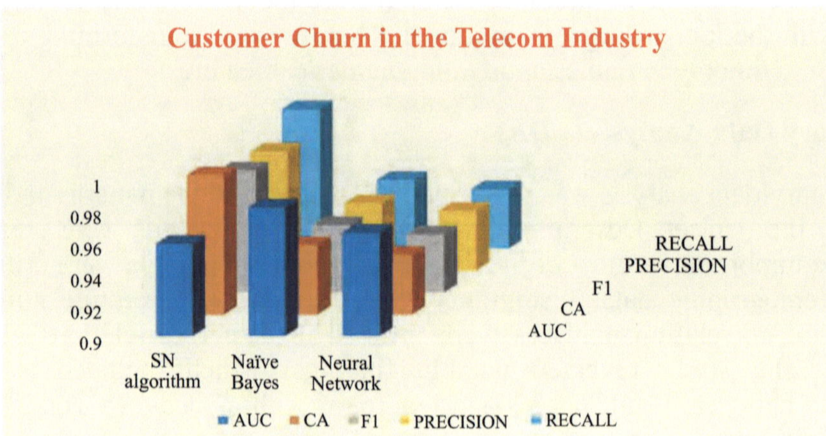

Fig. (2). Overview of the results obtained.

RESULTS

The attributes encompassed demographic information, utilized services, and expenditure patterns, all of which were crucial for constructing the prediction model. The research was conducted using the R programming language, utilizing its powerful packages for data preprocessing, visualization (Fig. **3**), and modeling.

Data Analysis

The exploratory data analysis (EDA) uncovered a well-balanced distribution in terms of gender and relationship status among the sample, with a minority consisting of senior citizens and individuals with dependents. The visualizations highlighted varied customer preferences, showing a clear preference for fiber optic over DSL for internet services (Fig. 4). This stage of analysis helped identify key patterns and trends in the data, providing a solid foundation for further investigation.

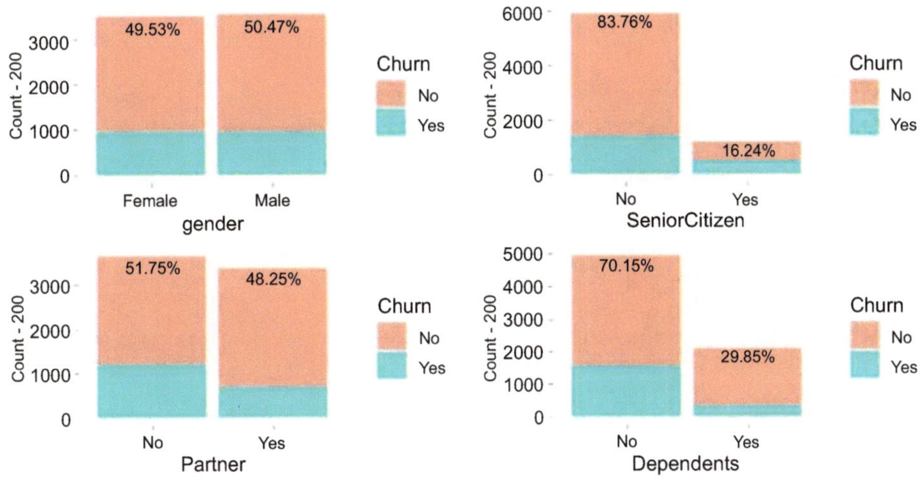

Fig. (3). These demographic plots show that the sample is evenly split across gender and partner status. A minority of the sample are senior citizens, and a minority have dependents.

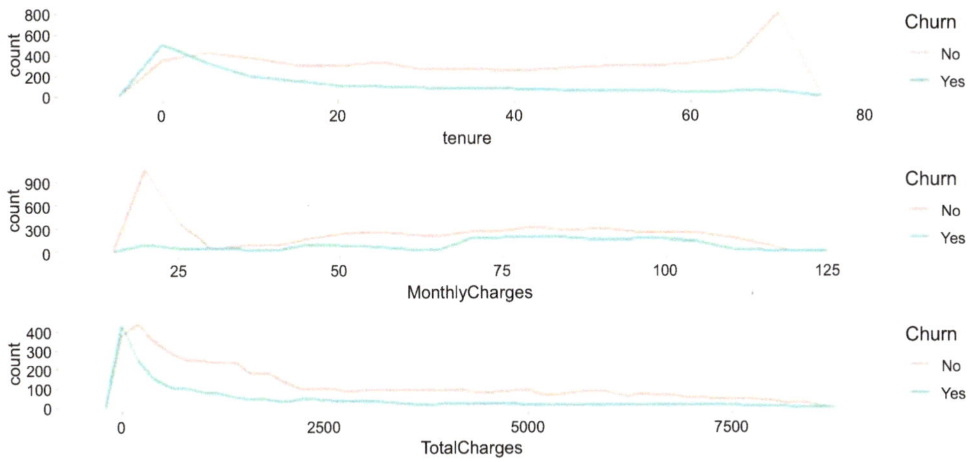

Fig. (4). The tenure variable is stacked at the tails, so a large proportion of customers have either had the shortest (1 month) or longest (72 months) tenure.

Data Preprocessing

Data preprocessing refers to the process of cleaning and transforming raw data into a format suitable for analysis. Preparing the data involved purging and reformatting to optimize the model's performance. The dataset was simplified by recoding dependent variables, such as Multiple-Lines and Online-Security. Additionally, extraneous variables, such as customerID, were eliminated to streamline the modeling process. The data was split into training and test subsets in a 70:30 ratio to assess the model's performance.

Construction and Assessment of Models

The following modeling techniques were used:

Neural Networks

The number of neurons in the hidden layers of the neural network was maximized to 100, using the ReLU activation function and the Adam solver algorithm, with regularization applied (alpha = 0.0002).

Gradient Boosting

We used a method called XGBoost, with the learning rate fixed at 0.3 and the number of trees set to 100 to ensure replicable training. Regularization (lambda) was set to 1, and the maximum depth of individual trees was limited to 4.

Naïve Bayes

Naïve Bayes was implemented using the Gaussian Naïve Bayes algorithm. The var_smoothing hyperparameter specifies the amount of variance smoothing to apply. A small alpha value was used as a smoothing parameter, added to the count of each feature to avoid zero probabilities.

Principal Discoveries

Contract Type

Customers on a month-to-month basis exhibited a higher likelihood of churning compared to those with longer-term contracts.

Internet Service

Customers utilizing fiber optic internet exhibited a higher propensity to discontinue their service compared to those utilizing DSL or having no internet access.

The study found that longer tenures and higher overall costs were associated with lower churn rates, suggesting that customer loyalty and financial stability are key determinants of customer retention.

Summary

The analysis found that regression was the most effective model for forecasting customer attrition, achieving the highest accuracy rate. The study emphasized key factors that telecom businesses can prioritize to reduce customer turnover. By understanding and leveraging these indicators, firms can strengthen their customer retention strategies, ultimately improving business performance. Future research could explore the integration of advanced machine learning algorithms and larger datasets to enhance prediction accuracy and improve model robustness. A comparative analysis of the models is presented below Table **1**, highlighting differences in AUC, accuracy, F1 score, precision, and recall values among the models used.

Table 1. The table demonstrates the values obtained after executing various models.

Model	AUC	CA	F1	Precision	Recall
SN Algorithm (SVM + Na¨ıve Bayes)	0.959	0.989	0.978	0.976	0.989
Na¨ıve Bayes	0.982	0.944	0.943	0.943	0.944
Neural Network	0.966	0.938	0.937	0.938	0.938

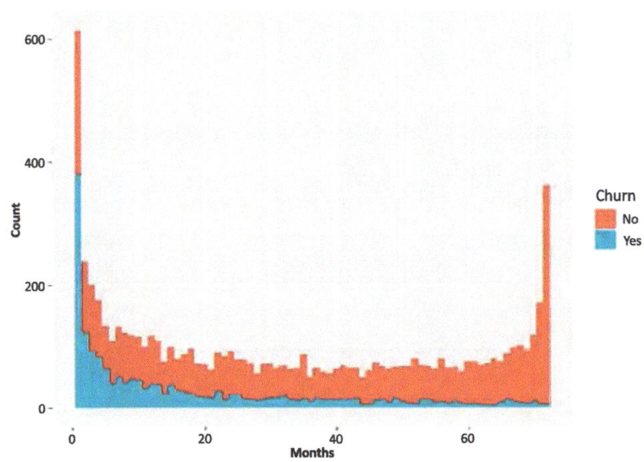

Fig. (5). The length of time as a customer decreases the likelihood of churn. There is a large spike at 1 month, indicating that a large portion of customers will leave after just one month of service.

CONCLUSION

In the communications industry, it is crucial to prioritize client retention because customers are likely to switch to competitors that offer more appealing programs. This study utilized an extensive dataset obtained from a telecoms business to investigate the prediction of customer attrition. The objective of the investigation was to identify the primary elements that influence customer turnover and develop prediction models to assist in creating strategic interventions for customer retention.

The study commenced with a comprehensive Exploratory Data Analysis (EDA) to reveal patterns and trends inherent in the data. Subsequently, data pretreatment procedures were conducted, encompassing tasks such as cleansing, recoding, and converting data to ensure its compliance with various modeling methodologies. The analysis included demographic characteristics, service utilization measures, and financial information to forecast customer attrition. The study included a few statistical techniques. Every strategy yielded valuable information about the key factors that predict client churn. The decision tree model highlighted the importance of contract type, duration, and internet service type, indicating that customers with month-to-month contracts and those using fiber-optic internet were more likely to churn. The Random Forest model validated these findings and highlighted the importance of total charges as a predictor.

Regression proved to be the most efficient technique, attaining the highest accuracy rate. Although regression is less complex compared to other models, it offers a balanced approach with a lower risk of false positives. The analysis highlighted tenure, contract status, and total charges as crucial determinants. Customers who have been with the company longer and have higher overall charges are less likely to discontinue the service, suggesting a positive relationship between customer loyalty and the amount invested in the service. The findings also imply that consumers with month-to-month contracts are more likely to churn, indicating that marketing longer-term contracts could be an effective strategy to improve retention. Furthermore, customers with fiber optic internet service exhibited higher churn rates, suggesting that this trend might be influenced by factors, such as service quality or competition. Understanding these characteristics enables telecommunication businesses to tailor their retention strategies more precisely.

Overall, the study highlighted the effectiveness of regression in accurately forecasting customer attrition in the telecom industry, revealing that most customers tend to leave the company within the first month of joining [Fig. 5]. Telecommunication firms can enhance customer loyalty and reduce churn by identifying and evaluating key churn indicators, allowing them to implement targeted solutions. The results support the recommendation to strategically promote longer-term contracts and improve service offerings to maintain client loyalty. Future research could build on these findings by incorporating additional variables and exploring more advanced machine learning methods to improve prediction accuracy. This approach ensures that customer attrition is addressed proactively, thereby reducing financial losses and increasing customer satisfaction.

REFERENCES

[1] J. Smith, and J. Doe, "A machine learning approach to customer churn prediction in the telecommunications sector", *Journal of Telecommunications and Information Technology,* vol. 12, no. 3, pp. 245-258, 2023.

[2] M-J. Lee, "Comparative analysis of machine learning algorithms for churn prediction in telecom networks", *Proceedings of the International Conference on Machine Learning and Data Science,* pp. 101-108, 2022.

[3] M. Garcia, "Leveraging machine learning for customer retention in the telecommunications industry", *Telecom Business Review,* vol. 47, no. 2, pp. 32-38, 2020.

[4] W. Chen, and D. Zhang, "Optimizing customer lifetime value in tele- com through machine learning-based churn prediction", *IEEE Trans. Knowl. Data Eng.,* vol. 33, no. 7, pp. 1789-1802, 2021.

[5] A. Khan, "The role of artificial intelligence in telecom customer churn management: A review", *Sensors (Basel),* vol. 22, no. 12, pp. 1-20, 2022. [doi: 10.3390/s22124700]. [PMID: 35746352]

[6] B. Johnson, *Using Machine Learning to Identify At-Risk Customers: A Churn Prediction Model for Telecom Companies.,* 2024.

[7] P. Patel, "Data-Driven Customer Retention: Applying machine learning to predict churn in telecom", *International Conference on Big Data and Analytics,* 2023.

[8] W. Miller, "Taming the Churn Monster: How machine learning can save telecom companies money", *Telecom Technology Magazine,* vol. 18, no. 4, pp. 52-58, 2021.

[9] S. Brown, "Predicting the Future: Machine learning for customer retention in telecom", *Telecom Innovation Journal,* vol. 7, no. 1, pp. 11-18, 2022.

[10] D. Williams, "Don't Let Them Go! using machine learning to keep telecom customers happy", *Customer Retention Strategies Journal,* vol. 3, no. 2, pp. 78-84, 2020.

[11] A. Gupta, and S. Singh, "A deep learning approach for churn pre- diction in telecommunication industry", *Int. J. Comput. Appl.,* vol. 187, no. 8, pp. 15-22, 2022. [doi: 10.5120/ijca2022918141]. [http://dx.doi.org/10.5120/ijca2017913660]

[12] J. Yang, "Cost-Sensitive Machine Learning for Customer Churn Predic- tion in Telecom Services", *IEEE Trans. Serv. Comput.,* vol. 15, no. 1, pp. 231-244, 2022. [doi: 10.1109/TSC.2019.2952222].

[13] A. Manzoor, M. Atif Qureshi, E. Kidney, and L. Longo, "A Review on Machine Learning Methods for Customer Churn Prediction and Recommendations for Business Practitioners", *IEEE Access,* vol. 12, pp. 70434-70463, 2024. [http://dx.doi.org/10.1109/ACCESS.2024.3402092]

[14] M. Szeląg, and R. Słowiński, "Explaining and predicting customer churn by monotonic rules induced from ordinal data", *Eur. J. Oper. Res.,* vol. 317, no. 2, pp. 414-424, 2024. [http://dx.doi.org/10.1016/j.ejor.2023.09.028]

[15] Nisha Gurung, "AI-Based Customer Churn Prediction Model for Business Markets in the USA: Exploring the Use of AI and Machine Learning Technologies in Preventing Customer Churn", *Journal of Computer Science and Technology Studies,* vol. 6, no. 2, pp. 19-29, 2024. [http://dx.doi.org/10.32996/jcsts.2024.6.2.3x]

[16] Z. Liu, P. Jiang, K.W. De Bock, J. Wang, L. Zhang, and X. Niu, "Extreme gradient boosting trees with efficient Bayesian optimization for profit-driven customer churn prediction", *Technol. Forecast. Soc. Change,* vol. 198, p. 122945, 2024. [http://dx.doi.org/10.1016/j.techfore.2023.122945]

[17] A. De Caigny, K.W. De Bock, and S. Verboven, "Hybrid black-box classification for customer churn prediction with segmented interpretability analysis", *Decis. Support Syst.,* vol. 181, p. 114217, 2024. [http://dx.doi.org/10.1016/j.dss.2024.114217]

[18] P. Jiang, Z. Liu, M.Z. Abedin, J. Wang, W. Yang, and Q. Dong, "Profit-driven weighted classifier with interpretable ability for customer churn prediction", *Omega,* vol. 125, p. 103034, 2024. [http://dx.doi.org/10.1016/j.omega.2024.103034]

CHAPTER 11

A Water Quality Prediction and Assessment Model using Machine Learning Classifiers

M. Vaishali[1,*], **M. Monika**[2], **M. Atish**[3] and **Lalitha**[4]

[1] *Department of Computer Science and Engineering, Maharishi Markandeshwar Deemed to be University, Ambala, India*

[2] *Department of Information Technology, Dwarkadas J Sanghvi College of Engineering, Navi Mumbai, India*

[3] *Department of Computer Science and Engineering, BMS Institute of Technology and Management, Banglore, India*

[4] *School of Advanced Sciences, VIT-AP University, Andhra Pradesh, India*

Abstract: Ensuring safe and clean water availability is vital to the health of not only human beings but also all species. Additionally, it is crucial for the sustainability of the environment. With the emergence of advanced technologies like machine learning, predictive models can significantly contribute to assessing and managing water quality. Current research proposes a methodology that predicts water quality using several machine learning classifiers on a dataset comprising diverse parameters, such as pH levels, dissolved oxygen, turbidity, and other pollutants, collected from multiple water sources. Initially, the data were preprocessed to remove missing values and outliers. Feature engineering was employed to identify the most relevant parameters that contribute to water quality. Several popular machine learning classifiers, including Random Forest, Support Vector Machines, Decision Trees, and XGBoost, were evaluated and compared for their performance in predicting water quality. The trained models were validated and tested using cross-validation techniques to ensure generalizability and resilience. The research findings demonstrated that the proposed method is effective in accurately forecasting water quality levels. The XGBoost, in particular, exhibited superior performance with high accuracy and minimal overfitting. Additionally, feature importance analysis revealed key factors influencing water quality, providing valuable insights for policymakers and environmentalists.

Keywords: Cross-validation, Machine learning classifiers, ROC curve, Water quality assessment.

* **Corresponding author M. Vaishali:** Department of Computer Science and Engineering, Maharishi Markandeshwar Deemed to be University, Ambala, India; E-mail: drvaishaliwadhwa@gmail.com

Suneeta Satpathy, Sachi Nandan Mohanty & Subhendu Kumar Pani (Eds.)

INTRODUCTION

The primary source of clean and safe drinking water for residential homes, businesses, and agriculture is groundwater, which is sourced from rivers, streams, dams, lakes, reservoirs, creeks, and wetlands [1]. Groundwater is severely impacted by pollution from unrestricted human activities, including industrialization and agriculture, despite its essential role in the survival of humans and other ecosystems, such as aquatic plants and wildlife. Nearly 90% of untreated sewage from underdeveloped nations enters water sources, according to the 2015 United Nations World Water Development Report [2]. In the meantime, approximately 300–400 megatons of waste from the sector are dumped into water sources annually, according to data from the United Nations Educational, Scientific, and Cultural Organization (UNESCO) [3].

Most nations, including India, also have high rates of uncontrolled sewage treatment or discharge from manufacturing and agriculture [4]. River pollution is primarily caused by aluminum, ammoniacal nitrogen (NH3-N), suspended solids (SS), cadmium, chloramine, and biochemical oxygen demand (BOD). High BOD can be caused by wastewater and inadequate sewage treatment from the industry and agro-based sectors. Meanwhile, elevated NH 3-N may be caused by animal husbandry and unchecked household sewage. Furthermore, inappropriate earthworks and land-clearing practices may affect elevated SS [5]. To identify which areas are contaminated and require treatment, ongoing data on the water quality of these river pollution hotspots are needed. The evaluation of Indian water quality has been extensively covered in previous research. However, a new, effective method to track pollution hotspots has been developed due to the growth and complexity of massive data from ambiguous water quality metrics. To effectively identify hotspots for river pollution, this study suggests employing a range of techniques and comparative analyses that utilize both supervised and unsupervised Machine Learning (ML) algorithms. ML algorithms and water quality have been the subject of several studies [6 - 8]. However, to our knowledge, there has not been a thorough comparison research in the literature on using ML with datasets used to assess river quality. Hence, seven popular algorithms, including supervised and unsupervised ML, were utilized to classify the pollution hotspots in the Indian water. These include Support Vector Classifiers (SVC), K-Nearest Neighbors (KNN), AdaBoost (AB), Naïve Bayes (NB), Logistic Regression (LR), Random Forest (RF), Gradient Boost (XG), and Decision Tree (DT).

Additionally, we adjusted and modified every parameter in ML algorithms to validate the accuracy of the algorithms. This procedure is essential to prevent underfitting and overfitting in ML algorithms [9]. Overall, the accuracy of the river pollution categorization was used to compare and evaluate each ML method in detail in this chapter. The preprocessing techniques and algorithm tuning procedures yielded the highest accuracy.

LITERATURE SURVEY

Different nations and organizations have produced many WQIs, such as the surface quality water index and groundwater quality index, to achieve particular objectives [10]. Many factors, such as (i) uncertainty concerns, (ii) model reliability, (iii) transparency, and (iv) model sensitivity, have been brought up against this technique. WQI model created a significant amount of uncertainty in the final score, according to recent studies (Juwana *et al.*, 2016; Sutadian *et al.*, 2018; Uddin *et al.*, 2021) [1, 10, 11]. Furthermore, several studies have recently shown that the water quality index model employs several classification techniques [12, 13]. It is recommended to use various qualitative measurements, such as "excellent," "good," "bad," "very bad," "poor," "marginal," "higher," and "lower," when interpreting the WQI score. As a result, many approaches offer a variety of interpretations for comparable water qualities, which adds to the significant uncertainty surrounding the accurate categorization of water quality.

Recent research indicates that, due to these problems, the water resource managers receive unclear information from the current WQI model, which makes it difficult for bodies to act promptly [1]. As previously stated, the ultimate objective of the WQI model is to utilize a classification system to categorize water quality [3] [6]. The outcome of the WQI model is often a numerical number, also referred to as an index score, that ranges from 0 to 100, where 100 represents "good" water quality and zero means "worst" water quality [7, 14]. Furthermore, according to numerous recent studies, a significant degree of uncertainty has been introduced into the WQI system due to the improper application of classification techniques. Therefore, for comparable groups of water quality indicators, the traditional classification algorithms may yield conflicting results in the final water quality evaluation [2]. Thus, in an earlier work [1, 11], the authors presented a universal classification scheme (see Table **3**) for assessing coastal and transitional water quality to minimize erroneous assessment and optimize the metaphor difficulties of existing classification systems. According to Uddin *et al.* [1], the universal scheme outcomes for water quality could be helpful in accurately representing various scenarios of water quality [1]. This chapter contains details about the classification scheme and development process. This study employed a grading scheme to assess India's current water quality status. Several ML classifier techniques were used to assess the WQI models' performance after obtaining water quality classifications. To reduce model uncertainty and improve water quality classification forecasts, sophisticated machine-learning methods have been applied extensively recently. Nowadays, most ML algorithms are created to solve binary classification problems [15, 16]. Because of this, several researchers [9, 17] have assessed WQI models using binary classes, even though the majority of WQI models in the literature recommend utilizing multiple classification schemes for evaluating water quality [18]. Using cutting-edge ML techniques, numerous researchers have created a variety of classifier methods to address the multiclass problem [19 - 21]. Support vector machines (SVM), Naïve Bayes (NB), random forests, decision trees, logistic regression, k-nearest neighbor (KNN), and gradient boosting (XGBoost) classifiers are among the most frequently utilized techniques for analyzing multiclass problems [22 - 24]. This study used six classifier algorithms—NB, SVM, KNN, DT, RF, and XGBoost—to predict multi-class coastal water quality categorization using the water quality index model. To the authors' knowledge, this study is the first to use a multi-class classifier technique to assess a WQI model's performance. The receiver operating characteristic (ROC) curves and confusion matrices have been used in current research to compare the model's sensitivity, accuracy, and efficiency in terms of multi-class classification, assessing the prediction performance of the classifier model(s). Additionally, this study utilized six classifier algorithms to generate the ROC curve(s), and the confusion matrix analysis technique was employed to evaluate the WQI model's performance in accurately classifying coastal water quality in India.

PROPOSED METHODOLOGY

A more accurate classification algorithm has been proposed in this chapter for the prediction and classification of the quality of coastal water. This classification mechanism has significantly reduced inconsistencies in the final water quality prediction. This is an improvement over the models described in the above studies. This research also aims to assess the performance of ML classifiers in evaluating transitional and coastal water quality, ensuring accurate categorization and prediction by incorporating the proposed classification mechanism. The objectives of the present research are as follows:

Table 1. Water quality classification assessment.

Classification Schemes	Range of WQI	Description
Good	80-100	Water is appropriate for all consumers.
Fair	50-79	Water is maintained with a bit of impairment.
Marginal	30-49	The water quality is not maintained, making it unsuitable for regular use.

(Table 1) cont.....

Classification Schemes	Range of WQI	Description
Poor	0-29	Quality is not safeguarded for many specific users, mainly aquatic life.

i. To develop an enhanced water quality classification mechanism that calculates the Water Quality Index (WQI) score for coastal water quality (Table **1**).
ii. To apply seven machine learning (ML) classifier models to the collected data and compare their performance using metrics such as Accuracy, Recall, Precision, and F1 score, in order to identify the best-performing predictive model.
iii. To evaluate the performance of the proposed WQI model using the ROC curve, validation curve, and confusion matrix across the different predictive models.

Dataset Collection

To conduct the research, data about water quality monitoring during 2019 were collected from a prevalent repository, *i.e.*, https://www.kaggle.com/datasets/anbarivan/indian-water-quality-data/discussion.

Table **2** details these indicators and their ideal range for coastal regions. To carry out our research, we identified the crucial indicators using the parameter importance graph shown in Fig. (**3**). Out of 10 samples, seven water quality indicators were found to be significant for the research.

Table 2. Water quality parameters.

Parameter	Standard Range for Indicators (Si)	Values of unit weights for each parameter (Wi)
Dissolved oxygen (DO)	10 mg/L	0.22
pH	8.5	0.26
Nitrate	45/100 mg/L	0.049
Biological oxygen demand (BOD)	5 mg/L	0.442
Conductivity	1000 μS/cm	0.002
Fecal coliform	100/100 mL	0.022
Total coliform	1000 mL	0.002

Applying ML Classifiers for Classification and Prediction

Based on the aforementioned classification mechanism, the most commonly used classifiers, such as NB, KNN, DT, SVM, and RF, are employed to predict the quality of various indicators. The accuracy obtained by these models is illustrated in Table **3**.

Table 3. Performance evaluation of various models

Model	Precision	Recall	F1 Score	Accuracy
RF	0.99	0.83	0.90	0.90
KNN	0.94	0.62	0.75	0.79
SVM	1.00	0.67	0.80	0.83
XGB	0.95	0.85	0.90	0.92
DT	0.83	0.83	0.83	0.83
NB	0.99	0.71	0.83	0.70

Naïve Bayes

Naïve Bayes, a scalable algorithm based on the principle of Bayes' theorem, is one of the most sought-after supervised ML algorithms, particularly for 2-way and m-way categorization of data. The number of parameters required for implementing NB must be proportional to the number of variables. It is a probabilistic approach that compares the likelihood of two identically independent variables. In the present research, the NB algorithm was only implemented considering these constraints.

KNN

KNN is also a prevalent binary classifier that categorizes data points based on their neighboring data points. KNN is the most reliable and fastest algorithm. One of the most potent features of KNN is that it can be easily integrated into any ML model, without requiring prior knowledge of the data point distribution. The KNN algorithm first calculates the distance between data points of the training set. Then, the new points are categorized based on the nearest point of the observed classes. The authors have used Euclidean distance to classify the data points into appropriate courses.

Support Vector Machine (SVM)

Support Vector Machine (SVM) is another popular algorithm for classification and regression is the Support Vector Machine (SVM). SVM was first proposed by Vapnik in 1995. Every data point is shown as a vector with n dimensions. The SVM model divides them into two discriminant classes using an n-1-dimensional separating hyperplane. The goal is to maximize the separation between the data points and the hyperplane. Studies show that SVM is as accurate as, or more accurate than, logistic regression in classifying diseases.

Decision Tree

A decision tree is the most straightforward algorithm based upon a top-down tree-based categorization method that determines the optimal classifier by calculating information gain and entropy. DT is a multi-class classification method that works well. We applied the DT model to the dataset we assumed for our WQI model to predict classes, *i.e.*, "Good, Fair, Marginal, and Poor." DT is most commonly used for an imbalanced dataset. Other merits of decision-tree-based prediction models include their efficiency, speed, and resistance to missing values. Moreover, DT-based models can effectively manage problem variables and data points, making them most suitable for short-term predictions.

Random Forest

Random Forest is an ensemble learning method that combines multiple base classifiers, typically decision trees, trained independently on different subsets of the training data. It uses the bagging technique, which involves drawing random samples with replacement. Each decision tree is trained on a different random subset of the data and a randomly selected subset of features. For the final prediction, Random Forest uses a majority voting mechanism (for classification) or averaging (for regression). The algorithm uses measures such as the Gini index and information gain to optimally select feature subsets at each split. By averaging the outputs of multiple decision trees, Random Forest reduces variance and mitigates the risk of overfitting.

Bagging, also known as bootstrap aggregating, involves creating multiple versions of the original dataset by applying combinations and repetitions. This technique reduces prediction variance by generating new training datasets from the original data. Random Forest (RF) leverages this approach to combine the advantages of decision trees with the robustness of ensemble learning. In traditional decision trees, each node is split based on the best feature among all available parameters. In contrast, Random Forest splits each node using the best feature from a randomly selected subset of parameters, introducing diversity among the trees and improving overall model performance.

XGBoost

The most popular ensemble learning strategy, "boosting," combines several weak classifier models. This method has been widely applied in ML research lately to discover possible models for classification and regression automatically. Furthermore, this algorithm is commonly used in many ML applications, such as incremental learning, class-balanced data, confidence estimate, feature selection, missing features, and error correction. Chen (2016) successfully proposed the gradient boost method, which has been upgraded and optimized into the XGBoost algorithm. Given its efficacy and versatility in practical applications, the XGBoost algorithm was used in this investigation to predict water quality classes.

Dataset Preparation and Preprocessing

The next step is preparing data for prediction modeling. As discussed above, the final dataset consists of 7 independent variables and four classifications - “good,” “fair,” “marginal,” and “poor.” After performing data preprocessing steps, such as handling missing values and normalization, seven ML classifiers were implemented on the trained data.

Evaluation Criteria

A confusion matrix (CM) is used to analyze the accuracy, sensitivity, and classification performance of prediction models. The four components that often make up a CM are true positives (TP), false negatives (FN), false positives (FP), and true negatives (TN). Accuracy, precision, sensitivity, F1 score, receiver operating characteristics (ROC), and area under the curve (AUC) were all used in this investigation. Metrics for model evaluation were calculated using the formula below; a projected classifier model's ROC curve shows the average sensitivity for all conceivable specificity values. The overall WQI likelihood scores (on the x-axis) and overall WQI value occurrence (on the y-axis) were considered in this study to create the ROC curve. The definition of accuracy is as follows:

$$Accuracy = \frac{TP+TN}{TP+TN+FP+FN} \tag{1}$$

The degree of similarity between the algorithm's predictions and the observations of the same categorization is another crucial metric for measuring precision. This is how the precision is expressed:

$$Precision = \frac{TP}{TP+FP} \tag{2}$$

Typically, the TP rate is related to the model's sensitivity or recall. It measures how often the algorithm correctly identifies positive instances in the dataset. False positive cases occur when the classifier labels observations as positive, even though their true class is negative. Sensitivity and F1-score are defined as follows:

$$Sensitivity = \frac{TP}{TP+FN} \quad \textbf{(3)}$$

$$F1\ Score = 2 \times \frac{Precision \times Sensitivity}{Precision+Sensitivity} \quad \textbf{(4)}$$

RESULTS AND DISCUSSION

The performance of the predictive classifier model is assessed using the ROC curve technique. The ROC curve evaluates model efficiency by examining the area under the curve (AUC), a widely used metric in machine learning. The AUC reflects the model's ability to discriminate between classes by summarizing the trade-off between sensitivity and specificity. Essentially, it quantifies the overall accuracy of the predicted classes. In this study, the AUC value was used to assess the classifiers' predictive ability by comparing predicted water quality classes against the actual classes to determine classification accuracy. According to Walter (2005), the AUC value ranges from 0 (no discriminative capability) to 1 (perfect discrimination). The performance of the classifier can be interpreted using five key thresholds:

i. AUC ≥ 0.9: The classifier model can accurately classify all classes.
ii. AUC = 0.8 – 0.9: The classifier's likelihood of correctly classifying an object is high.
iii. AUC = 0.70 – 0.79: A few misclassifications are acceptable for classifier performance.
iv. AUC = 0.5–0.7: The classifier makes a larger percentage of incorrect classifications.
v. AUC < 0.5 indicates that the model lacks discriminating capabilities, meaning the classifier cannot discern between different classes.

ROC Curve

In the present research, the ROC curves were obtained from seven ML algorithms. These are presented in Fig. (**1**).

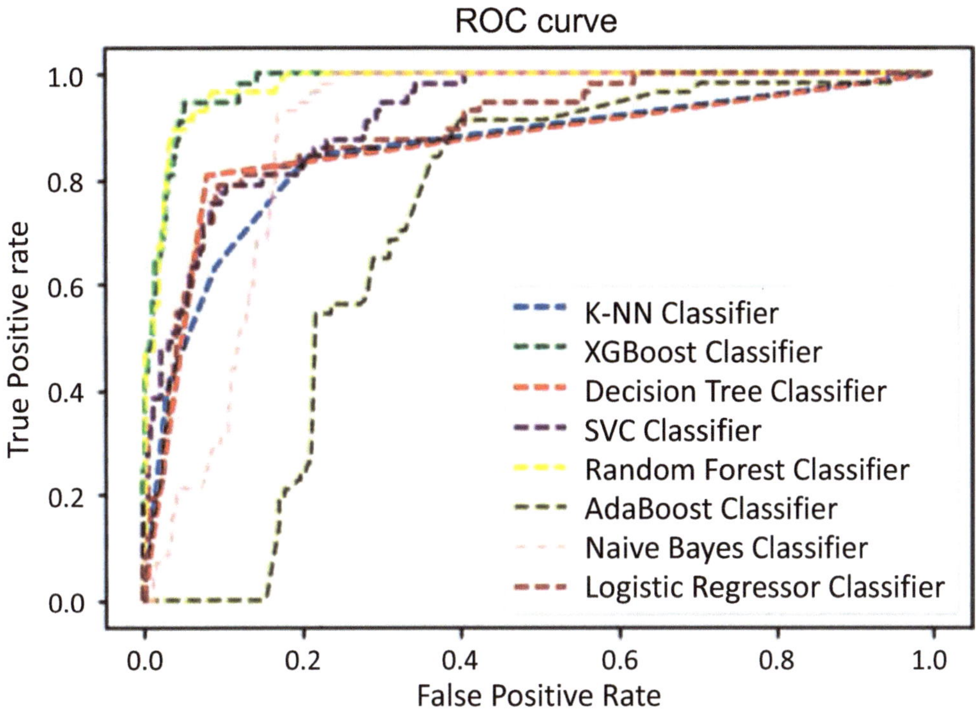

Fig. (1). ROC curves of seven ML algorithms for the multi-class categorization of water quality.

Statistical Summary

The Z statistics of the water quality indicators in the gathered dataset are shown in Fig. (**2**). Except for DIN, TON, TRAN, MRP, and DOX, the significant indicators of water quality were determined to be within the criteria for coastal water quality. As illustrated in Fig. (**4**), the BOD5, MRP, and pH were determined to have normal data distributions, while most indicators—aside from the TRAN and TEMP—showed data distribution patterns with positive skew. Except for MRP, the Z scores of most quality parameters were determined to be between -2 and + 2. According to the Z statistics results, most pointers in the dataset acquired without BOD5, CHL, and pH considerably influenced water quality. All indicators, except these, showed the presence of data outliers.

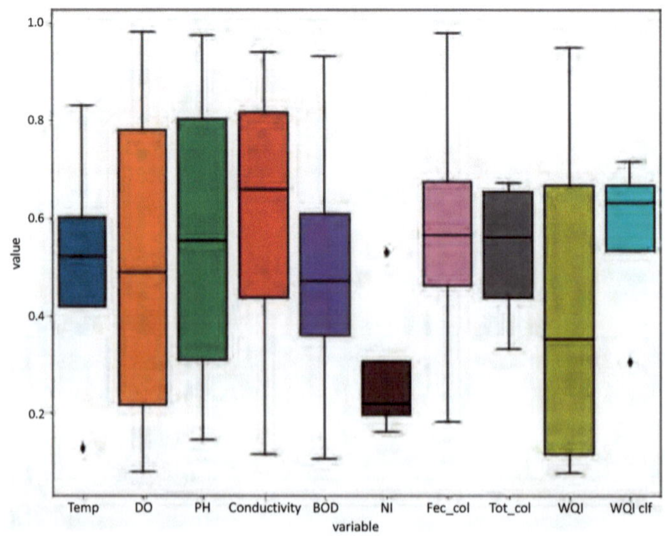

Fig. (2). Z-statistics of WQI parameters.

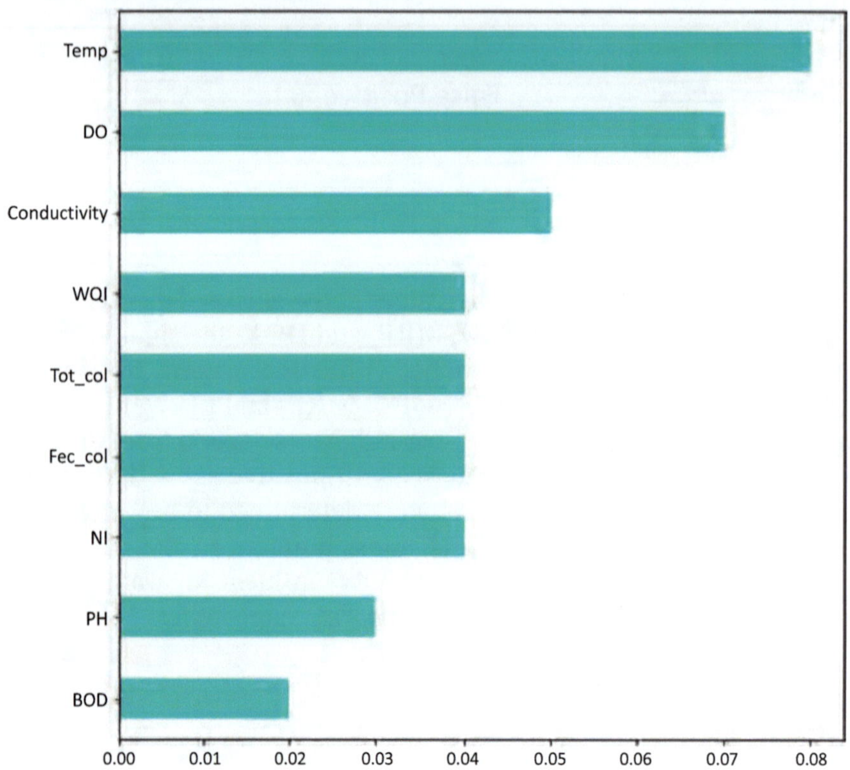

Fig. (3). Relative importance of attributes of water quality.

Selecting Essential Predictors for the Classifier Models

Using the XGBoost algorithm, the current study conducted a relative importance analysis to determine the impact of water quality indicators on classification. We found that TRAN, CHL, AMN, and MRP positively affect water quality. The critical indicators and their corresponding ranks are displayed in Fig. (**3**). Furthermore, this study investigated the correlation between water quality indicators across different water quality classes and identified their relationships using the parallel coordinates technique, also shown in Fig. (**3**). The figure illustrates that TRAN and CHL significantly influence marginal water quality, while DIN has a greater impact on poor water quality.

Confusion matrix of WQI model

The current study compared the effectiveness of 7 ML classifiers to determine which algorithm performs the best in terms of accurate categorization. For the unbalanced dataset, the classifiers' performance was assessed using five validation measures, including the confusion matrices: accuracy, precision, sensitivity, specificity, and F1 score. The confusion matrices for each of the four predictive classifier models are displayed in Fig. (**4**). In this investigation, classification was predicted using 1000 data points that fell into four categories: "good," "fair," "marginal," and "poor."

The confusion matrix results for the WQI models examined for multi-class water quality categorization across different Indian states indicate that XGBoost performed best in accurately predicting water quality categories. The performance metrics of the various classifiers used to identify water types are presented in Table **3**. This table shows that the XGBoost classifier achieved the highest accuracy among the models tested. Therefore, the WQI model's water quality predictions are most effectively made using the XGBoost classifier.

The prediction performance of the four chosen algorithms is illustrated in Fig. (**5**). According to the results, XGBoost outperformed the other techniques across all five factors. Specifically, it demonstrated exceptional performance for the "good" to "marginal" classes, achieving an approximately 96% F1 performance score for the "poor" class in terms of precision and sensitivity.

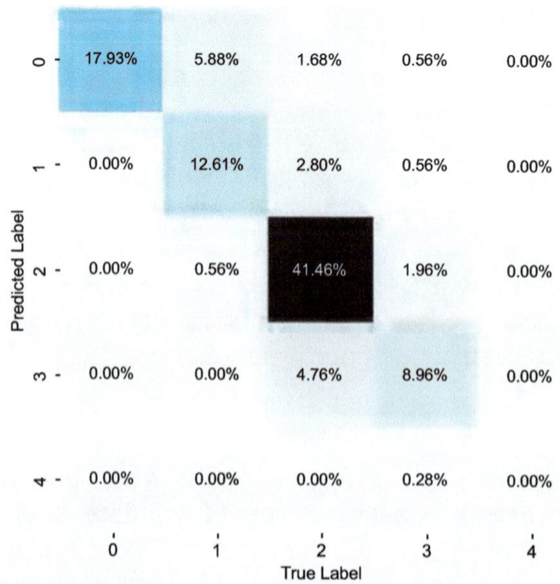

Fig. (4). Confusion Matrix of WQI Model.

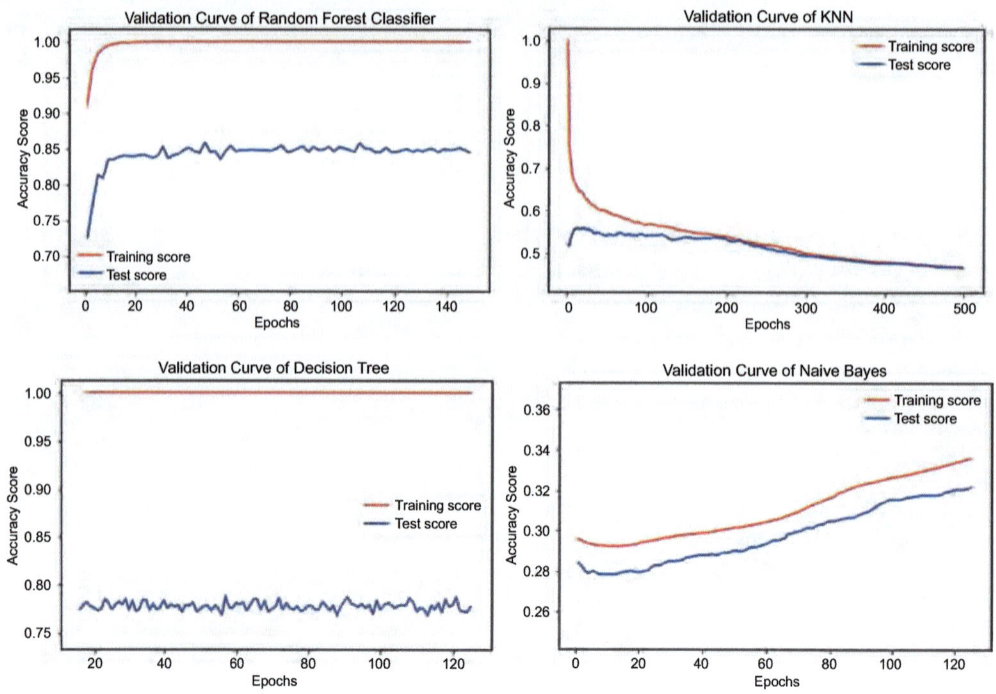

Fig. (5). Validation curve of various classifiers.

Various classifiers are implemented on the dataset and tested, with accuracy values recorded. It is found that the voting classifier is the best approach.

CONCLUSION AND FUTURE SCOPE

The primary goals of this study were to develop a framework for evaluating the performance of the WQI model and to accurately classify coastal water quality. The best machine-learning classifier algorithm for predicting water quality classes was identified by comparing four techniques. This study evaluated 7 ML models. It was shown that the XGBoost algorithm and KNN performed better in correctly classifying water quality. Except for the "poor" classes, XGBoost accurately categorized 100% of the water quality classifications. The quality of water in both models was classified into four classes: "Good," "Fair," "Marginal," and "Poor." To the best of our knowledge, however, this work offers the first thorough method for assessing the effectiveness of WQI model(s) using a new classification system for the multi-class categorization of coastal water quality for the Indian water under investigation. Furthermore, the study's findings may help classify water quality accurately, thereby enhancing the WQI model's accuracy, reliability, and transparency through the accurate classification of coastal water quality. This study's major flaw was that it overlooked the temporal fluctuations of the water quality indices in Indian water. Additional research should be conducted using different predictive classifier algorithms to evaluate the temporal resolution of indicators and assess the performance of the WQI model(s). Notwithstanding its limitations, the results of this study may help mitigate the risk of model uncertainty resulting from incorrect classification, thereby offering valuable insights for scholars, policymakers, and water research professionals.

REFERENCES

[1] A.M. Nayar, Z. Attar, S. Kachwala, T. Biswas, and S.K. Wagh, "Dost-mental health assistant chatbot", 2022 5th International Conference on Advances in Science and Technology (ICAST), Mumbai, India, 2022, pp. 252-257. [http://dx.doi.org/10.1109/ICAST55766.2022.10039576]

[2] M.G. Uddin, S. Nash, A. Rahman, and A.I. Olbert, "A comprehensive method for improvement of water quality index (WQI) models for coastal water quality assessment", *Water Res.,* vol. 219, p. 118532, 2022. [http://dx.doi.org/10.1016/j.watres.2022.118532] [PMID: 35533623]

[3] R. Abrahão, M. Carvalho, W.R. Da Silva Jr, T. Machado, C. Gadelha, and M. Hernandez, "Use of index analysis to evaluate the water quality of a stream receiving industrial effluents", *Water S.A.,* vol. 33, no. 4, pp. 459-466, 2010. [http://dx.doi.org/10.4314/wsa.v33i4.52940]

[4] M. Najafzadeh, F. Homaei, and H. Farhadi, "Reliability assessment of water quality index based on guidelines of national sanitation foundation in natural streams: integration of remote sensing and data-driven models", *Artif. Intell. Rev.,* vol. 54, no. 6, pp. 4619-4651, 2021. [http://dx.doi.org/10.1007/s10462-021-10007-1]

[5] S. Bedi, A. Samal, C. Ray, and D. Snow, "Comparative evaluation of machine learning models for groundwater quality assessment", *Environ. Monit. Assess.,* vol. 192, no. 12, p. 776, 2020. [http://dx.doi.org/10.1007/s10661-020-08695-3] [PMID: 33219864]

[6] S. Gupta, and S.K. Gupta, "A critical review on water quality index tool: Genesis, evolution and future directions", *Ecol. Inform.,* vol. 63, p. 101299, 2021. [http://dx.doi.org/10.1016/j.ecoinf.2021.101299]

[7] N.A. Suwadi, M. Derbali, N.S. Sani, M.C. Lam, H. Arshad, I. Khan, and K-I. Kim, "An optimized approach for predicting water quality features based on machine learning", *Wirel. Commun. Mob. Comput.,* vol. 2022, pp. 1-20, 2022. [http://dx.doi.org/10.1155/2022/3397972]

[8] N. Nasir, A. Kansal, O. Alshaltone, F. Barneih, M. Sameer, A. Shanableh, and A. Al-Shamma'a, "Water quality classification using machine learning algorithms", *J. Water Process Eng.,* vol. 48, p. 102920, 2022. [http://dx.doi.org/10.1016/j.jwpe.2022.102920]

[9] N. H. A. Malek, W. F. W. Yaacob, S. A. M. Nasir, and N. Shaadan, *Prediction of Water Quality Classification of the Kelantan River Basin, Malaysia, Using Ma-chine Learning Techniques,* 2022. [http://dx.doi.org/10.3390/w14071067]

[10] D.T. Bui, K. Khosravi, J. Tiefenbacher, H. Nguyen, and N. Kazakis, "Improving prediction of water quality indices using novel hybrid machine-learning algorithms", *Sci. Total Environ.,* vol. 721, p. 137612, 2020. [http://dx.doi.org/10.1016/j.scitotenv.2020.137612] [PMID: 32169637]

[11] M.G. Uddin, S. Nash, A. Rahman, and A.I. Olbert, "Performance analysis of the water quality index model for predicting water state using machine learning techniques", *Process Saf. Environ. Prot.,* vol. 169, pp. 808-828, 2023. [http://dx.doi.org/10.1016/j.psep.2022.11.073]

[12] M.G. Uddin, S. Nash, A. Rahman, and A.I. Olbert, "A novel approach for estimating and predicting uncertainty in water quality index model using machine learning approaches", *Water Res.,* vol. 229, p. 119422, 2023. [http://dx.doi.org/10.1016/j.watres.2022.119422] [PMID: 36459893]

[13] M. S. Islam Khan, N. Islam, J. Uddin, S. Islam, and M. K. Nasir, *Water quality prediction and classification based on principal component regression and gradient boosting classifier approach,* 2022. [http://dx.doi.org/10.1016/j.jksuci.2021.06.003]

[14] N. Zamri, "A comparison of unsupervised and supervised machine learning algorithms to predict water pollutions", In: *Procedia Computer Science* Elsevier B.V., 2022, pp. 172-179. [http://dx.doi.org/10.1016/j.procs.2022.08.021]

[15] V. Piccialli, and M. Sciandrone, "Nonlinear optimization and support vector machines", *Ann. Oper. Res.,* vol. 314, no. 1, pp. 15-47, 2022. [http://dx.doi.org/10.1007/s10479-022-04655-x]

[16] N. H. A. Malek, W. F. W. Yaacob, S. A. M. Nasir, and N. Shaadan, *Prediction of Water Quality Classification of the Kelantan River Basin, Malaysia, Using Ma-chine Learning Techniques,* 2022. [http://dx.doi.org/10.3390/w14071067]

[17] M. Ahmed, R. Mumtaz, and S.M. Hassan Zaidi, "Analysis of water quality indices and machine learning techniques for rating water pollution: a case study of Rawal Dam, Pakistan", *Water Sci. Technol. Water Supply,* vol. 21, no. 6, pp. 3225-3250, 2021. [http://dx.doi.org/10.2166/ws.2021.082]

[18] A. El Bilali, A. Taleb, and Y. Brouziyne, "Groundwater quality forecasting using machine learning algorithms for irrigation purposes", *Agric. Water Manage.,* vol. 245, p. 106625, 2021.

[http://dx.doi.org/10.1016/j.agwat.2020.106625]

[19] S. Singha, S. Pasupuleti, S.S. Singha, R. Singh, and S. Kumar, "Prediction of groundwater quality using efficient machine learning technique", *Chemosphere,* vol. 276, p. 130265, 2021. [http://dx.doi.org/10.1016/j.chemosphere.2021.130265] [PMID: 34088106]

[20] S.B.H.S. Asadollah, A. Sharafati, D. Motta, and Z.M. Yaseen, "River water quality index prediction and uncertainty analysis: A comparative study of machine learning models", *J. Environ. Chem. Eng.,* vol. 9, no. 1, p. 104599, 2021. [http://dx.doi.org/10.1016/j.jece.2020.104599]

[21] A. Nouraki, M. Alavi, M. Golabi, and M. Albaji, "Prediction of water quality parameters using machine learning models: a case study of the Karun River, Iran", *Environ. Sci. Pollut. Res. Int.,* vol. 28, no. 40, pp. 57060-57072, 2021. [http://dx.doi.org/10.1007/s11356-021-14560-8] [PMID: 34081285]

[22] V. Mehta, N. Batra, Poonam, S. Goyal, A. Kaur, K.V. Dudekula, and G.J. Victor, "Machine Learning based Exploratory Data Analysis (EDA) and Diagnosis of Chronic Kidney Disease (CKD)", *EAI Endorsed Trans. Pervasive Health Technol.,* vol. 10, 2024. [http://dx.doi.org/10.4108/eetpht.10.5512]

[23] N. Sharma, J. Dev, M. Mangla, V.M. Wadhwa, S.N. Mohanty, and D. Kakkar, "A heterogeneous ensemble forecasting model for disease prediction", *New Gener. Comput.,* vol. 39, no. 3-4, pp. 701-715, 2021. [http://dx.doi.org/10.1007/s00354-020-00119-7] [PMID: 33424081]

[24] V. Mehta, M. Mangla, N. Sharma, M. Rakhra, T. Choudhury, and G. Rama Krishna, "Machine learning based assessment and predictive analysis of in-vitro fertilization success rate", *EAI Endorsed Trans. Pervasive Health Technol.,* vol. 10, 2024. [http://dx.doi.org/10.4108/eetpht.10.5511]

CHAPTER 12

Synapsys – A Tool for Effective Video Summarization

S. Divya Meena[1], B.V. Chandrahaas[1], C. Sai Janakiram[1], J.L.B. Krishna[1], M. Rushane[1] and **J. Sheela[1,*]**

[1] *School of Computer Science & Engineering, VIT-AP University, Amaravati, Andhra Pradesh, India*

Abstract: In the 21st century, the internet has become a ubiquitous platform for exchanging information, contributing to an extreme daily influx of approximately 2.5 Exabytes of data. Statistics show that a substantial 1/5th of the data transmitted over the internet comprises videos. With 30,000 hrs. of videos uploaded on YouTube every hour, the need to evaluate and derive valuable insights from this massive amount of video data becomes apparent in a dynamic landscape. Particularly for professionals engaged in internet exploration for various purposes, such as prominent data analysts, data scientists, and cybercrime analysts, it becomes apparent that they need to have an understanding of it. Though traditional machine learning-based systems exist, they have become obsolete due to modern times' issues, including decreased performance, greater computing complexity, and resource-intensive processes. GenAI has effectively addressed these problems, which is considered a transformative force with enhanced accuracy and semantic comprehension. This chapter focusses on the task of video summarization, which has been poised to supplant antiquated methodologies with the cutting-edge capabilities of GenAI. We have also applied the concepts of hybridization and fine-tuning of models in machine learning to create an improved version of generative models. The proposed methodology has demonstrated better performance in obtaining information, understanding computational complexity, *etc.*

Keywords: Abstractive summarization, Extractive summarization, Generative AI, Video summarization.

INTRODUCTION

Video summarization has become prevalent in the 21st century due to the inflow of data over the internet, not only for gaining valuable insights from videos, but every person has unique needs with the video content. Also, the versatility of the format of the videos makes them even more popular, and psychologically, people

* **Corresponding author J. Sheela:** School of Computer Science & Engineering, VIT-AP University, Amaravati, Andhra Pradesh, India; E-mail: Sheela.j@vitap.ac.in

Suneeta Satpathy, Sachi Nandan Mohanty & Subhendu Kumar Pani (Eds.)

understand the content more accurately when it has been provided to them in the form of videos. Considering the above scenario, we can understand the need for video summarization, which helps the users gain insights without watching the entire content. This enables the respective platforms not only to reduce the rate of user interaction but also allow users to consume information in a very minimal amount of time. Now, the question of where this task of video summarization can be found useful often arises. The state-of-the-art video summarization can be used by platforms that host video content, commoners who consume video content, and students who usually rely on video content for knowledge and information. Also, the current hot topic of 2023 is TikTok resumes, where job applicants make a short video about themselves and forward it to the companies. In this case, multinational companies can also use such summarizer systems to screen candidates more quickly and efficiently. From the above context, we can see that these systems have a broader scope. A video summary can be understood as a concise paragraph of the entire video, highlighting the key information mentioned in it. The current system provides such a service with the help of generative artificial intelligence models, which consist of attention mechanisms as their backbone and can be deployed in scenarios where a video consists of background audio and both contexts are overlapped.

Related Work

Video summarization has witnessed significant developments, encompassing various methodologies and technologies. This section provides a comprehensive overview of related works, highlighting key studies, methodologies, and technological advancements that have contributed to the current state of the art. Early research on video summarization focused on extracting keyframes to condense video content effectively. Seminal works [1 - 4] introduced foundational concepts in keyframe-based summarization, laying the groundwork for subsequent developments. In a paper [5], this approach was extended by incorporating content-based retrieval techniques, enabling more nuanced video analysis. Another study [6] explored the application of visual saliency for summarizing soccer highlights, introducing the idea of leveraging visual attention to identify key events.

Divya et al.

A study [7] proposed a hierarchical approach to story-based video summarization, emphasizing the importance of organizing content at different levels of granularity. The advent of deep learning techniques has revolutionized video summarization. A study [8] surveyed the utilization of deep learning for image and video description, showcasing the potential of neural networks in understanding complex visual information. Another paper [9] introduced "BertSum," a model that utilizes pre-trained encoders like BERT for abstractive text summarization, influencing the text summarization component of our proposed Video Summarization System. Another study [10] proposed a multimodal approach integrating vision and text information for unsupervised video summarization. Their work highlights the significance of combining audio-visual cues and textual content to create more comprehensive and contextually rich summaries. Recent studies have explored the integration of transformer models for text summarization, showcasing their ability to generate coherent and contextually relevant summaries. A study [11] introduced GPT-3, a language model that exhibits few-shot learning capabilities, opening new possibilities for enhancing the text summarization aspect of our proposed Video Summarization System. The above works have inspired this study and aim to bring advancements in content-based video retrieval and the exploration of novel technologies for improving the overall efficiency and effectiveness of video summarization.

PROPOSED METHODOLOGY

Data Collection

Data collection becomes tedious while working with complex architectures like transformers, since they are well known for the errors in training cases with very little data. To avoid these problems, we have tried to collect as much data as possible and fine-tune the models with it so that the system performs better. We have used C4, HugeNews, and LibriSpeech datasets to fine-tune the models to create the system. C4 stands for Colossal Clean Crawled Corpus, a text dataset made by web crawling. It is a 750GB textual dataset for training LLMs and chatbots. HugeNews is a dataset comprising 1.5 billion news articles. The LibriSpeech corpus is a collection of approximately 1,000 hours of audiobooks that are a part of the LibriVox project. Most of the audiobooks come from Project Gutenberg. The training data is split into three partitions of 100hr, 360hr, and 500hr sets, while the dev and test data are divided into the 'clean' and 'other' categories, respectively.

Synapsys Architecture

After finding gaps in the previous literature mentioned above and getting exposure to the new technologies and methodologies available during the execution of this project, we came up with a method that integrates the power of generative AI models along with the traditional theory of hybrid models from machine learning. Above is the representation of the method in a flow diagram as shown in Fig. (**1**). Summarizing the flow diagram, initially, the user will approach our Synapsys platform with their already downloaded video or any other video available online and upload it to the system. Now, the video will be directed towards whispering to generate the transcript. This transcript is further processed towards our Pegasus model, an abstract pre-trained text summarization model that basically works on the architecture of transformers [12, 13]. After obtaining a summary from it, the text is further directed to another extractive text summarizer, which reduces the length of the summary while preserving the key information generated from it. This will be considered the final summary, and the same will be returned as an output to the user along with the actual transcript of the video, through which the user can compare both the references and proceed with the information they want.

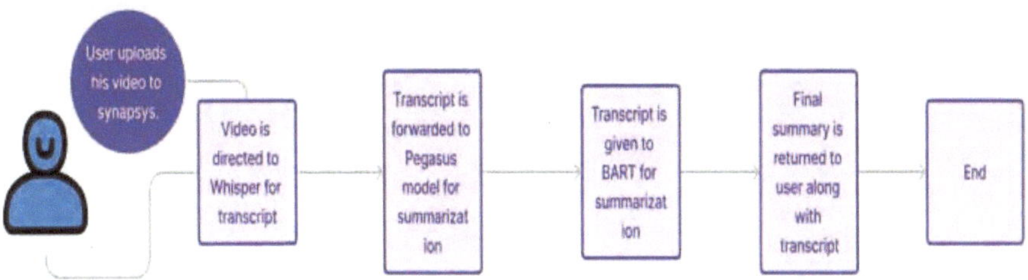

Fig. (1). Process flow diagram of methodology.

Whisper

Whisper is an automatic speech recognition (ASR) system developed by OpenAI. It is designed to convert spoken language into written text, making it a valuable tool for various applications, including transcription services, voice assistants, and other similar uses. The audio from the video is extracted and given to a whisper model for converting into text format. We can access Whisper through the API provided by OpenAI, which can be accessed for free. These developers can integrate high-end automatic speech recognition models into their applications. Whisper performs automatic voice recognition using a cutting-edge deep learning architecture [14, 15]. A large volume of multilingual and multitasking supervised web data trains the underlying model. For the model to acquire the subtleties and patterns of spoken language, it must be exposed to various audio data combined with appropriate transcriptions during the training process. Deep neural networks—more precisely, a blend of recurrent and convolutional neural networks—are a key component of Whisper's architecture, as shown in Fig. (**2**). Together, these networks enable the model to comprehend context and

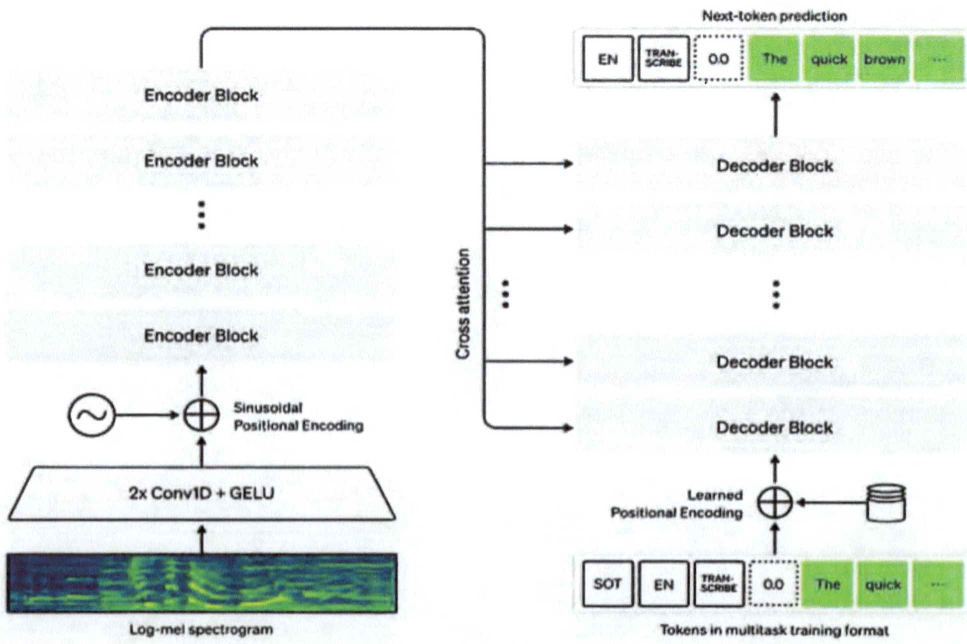

Fig. (2). Architecture of Whisper.

generate accurate transcriptions by capturing local and global relationships in the audio stream. Edge Splitting is done using the RandomLinkSplit() function, which randomly splits the entire graph into an 80:10:10 ratio, where 80 is used for training,10 for validation, and 10 for testing. Within training data, 70% is used for message passing, *i.e.*, actually training the model, and 30% for supervision, *i.e.*, ground truth values we want to compare our model's performance against. In training data, no opposing edges are added, opposing edges are non-existent edges added to the nodes. Negative edges are not added to training data because we want our graph to learn from positive feedback, *i.e.*, user interactions with the movie. While validating and testing, we add negative samples because it helps assess how well the model can genuinely distinguish between true and false links.

The system uses its trained neural network to interpret input when users upload an audio file or stream it to the Whisper API. The audio is transformed into a series of linguistic characteristics, which the model then examines to forecast the associated text. The model can handle different accents, languages, and speaking styles because of its capacity to generalize from its training data. The spoken content of the input audio is represented textually in Whisper's output. This transcript can be used for many things, such as typing oral lectures into text, adding closed captions to films, and enabling voice commands for apps. Whisper is a strong ASR system, but it is vital to remember that it might not be able to handle all circumstances, particularly ones with a lot of background noise, unusual dialects, or intricate technical jargon [16, 17]. To improve the accuracy of the generated transcripts, users should be aware of these possible issues and consider doing extra pre- or post-processing processes. In conclusion, OpenAI's Whisper is an advanced ASR system that uses deep learning to translate audible words into written text. Its architecture enables it to handle various audio inputs for multiple applications, and users can access it through OpenAI's API. Below is the Whisper framework, a significant combination of transformer models. We divided the audio into 30-second chunks and fed the stream to the model. We have fine-tuned the model with the "Librispeech" dataset for better efficiency. The transcript generated is further given to the Pegasus model for summarization.

Pegasus

Pegasus is a pre-trained abstractive summarizer that does not directly extract the important topics by any other traditional methods like ranking, but performs semantic understanding of the given text and then extracts key information from it. It is based on transformer architecture, similar to other models like GPT. Pegasus is built upon the transformer architecture, which includes attention mechanisms to capture long-range dependencies in the input text efficiently. Transformers are known for their parallelization capabilities and have been widely adopted in various natural language processing (NLP) tasks. Pegasus follows an encoder-decoder framework. The input document is encoded into a fixed-size representation by the encoder, and then the decoder generates the summary based on this representation. This architecture is fundamental to many sequence-to - sequence models, as shown in Fig. (**3**) [18 - 20].

Pegasus is pre-trained on a massive corpus of diverse text data. During pre-training, the model learns contextualized representations of words and sentences. After pre-training, the model can be fine-tuned on specific summarization tasks using datasets that include pairs of documents and their corresponding human-generated summaries. Pegasus is pre-trained on a massive corpus of diverse text data. During pre-training, the model learns contextualized representations of words and sentences. After pre-training, the model can be fine-tuned on specific summarization tasks using datasets that include pairs of documents and their corresponding human-generated summaries [21].

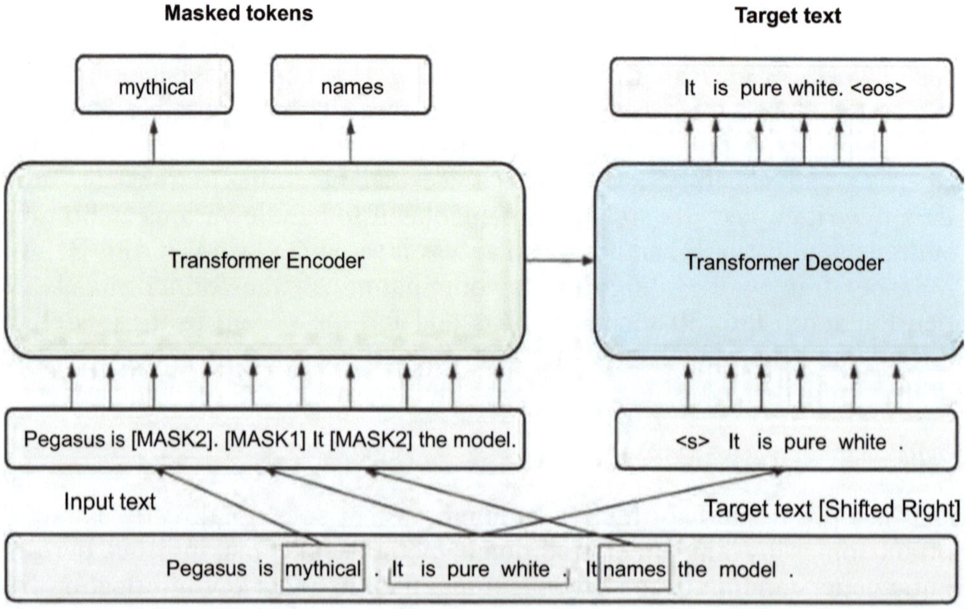

Fig. (3). Architecture of Pegasus.

Pegasus can be applied in various domains where summarization is valuable, such as news articles, research papers, and legal documents. Its ability to concisely capture a document's essence makes it useful for quickly understanding and conveying the key points of lengthy texts. We have fine-tuned the model on C4 and HugeNews datasets to enhance efficiency. Its architecture and a pictorial demonstration of how it summarizes a sentence and extracts key info are shown above in Fig. (**3**) [22].

RESULTS AND DISCUSSION

This section will discuss the results and scientific findings of the above system. Any software or system, in general, needs to undergo performance measures to ensure the system works fine with some benchmark results. Generally, for any Traditional Machine learning models, the performance metrics used to assess the system's performance would be Accuracy, Precision, Recall & F1-score. But when it comes to summarizing models and Natural Language Processing tasks, ROUGE scores are considered performance metrics. Since our system performs tasks of summarizing and we can only know that the generated summary is perfectly accurate by comparing the n-grams of the transcript generated and the summary generated, and this way we can figure out whether the model is performing fine. ROUGE stands for Recall-Oriented Understudy for Gisting Evaluation. It checks the number of overlaps in the n-grams of the summary generated and the Transcript obtained from the video. We have calculated the ROUGE-1, ROUGE-2, and ROUGE-L scores for the above system, as well as the BLEU scores. We have also compared them with existing studies on this problem statement or relevant fields. Table **1** below presents the experimental values.

Table 1. Comparative table of results b/w Synapsys and previous works.

Model	Rouge-1	Rouge-2	Rouge-L
BertSum	0.32	0.11	0.37
Roberta	0.39	0.17	0.32
BART	0.42	0.20	0.30
GPT-2	0.29	0.40	0.26
BERT	0.40	0.45	0.55
PEGASUS	0.45	0.22	0.36
SYNAPSYS	0.70	0.50	0.61

It can be seen that the overlap score was much higher for Synapsys (current system) than the previously available models. The models with a BLEU score greater than 50 can be considered good in terms of performance. We have achieved a score of 50.80 for the system; hence, based on all these findings, we can conclude that our system is reliable and performs well. The high ROUGE-1, ROUGE-2, and ROUGE-L scores imply a significant overlap between the n-grams of the generated summary and the transcribed content from the video. This indicates that the summarization system effectively captures and reproduces key content from the input, showcasing its ability to distill essential information. The ROUGE-L score, which measures the longest common subsequence, indicates that the system excels at maintaining contextual understanding and coherence between the original transcript and the generated summary. This is crucial for ensuring that the summarized output retains the overall meaning of the source content.

BLEU scores greater than 50 are considered good performers. This benchmark achievement signifies that the system excels in generating summaries that closely align with human reference summaries, demonstrating high accuracy and relevance. The comparison with existing studies suggests that Synapsys outperform previous models in the field. The higher ROUGE scores and the notable BLEU score indicate that Synapsys is a reliable and advanced system for video summarization. The findings collectively support the conclusion that the Synapsys system is a robust and high-performing solution for video summarization. Its ability to generate summaries with substantial overlap, maintain contextual coherence, and achieve a commendable BLEU score positions it as a reliable tool in natural language processing. While the current evaluation showcases the effectiveness of Synapsys, future work could explore additional dimensions, such as user studies of subjective assessment and testing the system on diverse datasets. Continuous refinement and adaptation of the models based on user feedback and emerging technologies could enhance the system's performance and versatility in various real-world applications.

CONCLUSION AND FUTURE SCOPE

The journey of conceptualizing, designing, and developing the Video Summarization System has been a testament to the fusion of innovation, technology, and user-centricity. From its inception to the culmination of its primary objectives, this project marks a significant leap toward revolutionizing video comprehension and accessibility across diverse domains. The inception of this project stemmed from a visionary goal, to create a system that transcends the traditional viewing experience, empowering users to glean comprehensive insights from videos without navigating their entirety. The roadmap unfolded with meticulous planning, algorithmic research, and a commitment to harnessing cutting-edge technologies. The journey culminated in the realization of pivotal objectives. The system's hybrid architecture has emerged as a paragon of efficiency, particularly in YouTube videos and TikTok resumes, aligning seamlessly with the project's fundamental mottos. The transformative potential of this system has been elucidated, offering practical applications in academia, corporate recruitment, and content consumption. In education, the system's adaptability has surfaced as a game-changer. Students and educators can harness this tool to distill complex video lectures and assignments into concise summaries, optimizing learning outcomes and facilitating knowledge assimilation. The promise of streamlining educational content consumption is evident, bridging the gap between extensive lectures and time-constrained study sessions.

The quest for improvement is perpetuated with the pursuit of advanced models. Integrating these models with robust frameworks and deploying them in optimized environments such as Docker, is a testament to the project's commitment to enhancing efficiency and scalability. The system's evolution is not merely about functionality but a continuous pursuit of technological excellence. Creating a centralized database for summarized content signifies a pivotal step toward efficiency. This database streamlines access, reducing computation time for future searches of identical video content. Simultaneously, real-world

integration across diverse industries, such as entertainment, news dissemination, or academic research, becomes a cornerstone for the system's adaptability. User-centricity remains at the core. The envisioning of a user-friendly mobile application amplifies accessibility, ensuring seamless utilization across devices. However, the project's grandeur extends beyond solitary development; it calls for collaboration. Dockerization and the project's transition to an open-source platform beckon a community-driven approach, fostering collaboration and innovation through collective efforts. Providing a system that accurately captures the essence of video content marks a pivotal milestone. The hybrid architecture exhibits exceptional performance, particularly on platforms like YouTube and TikTok, aligning seamlessly with the project's primary objectives. Expanding the system's utility for students and lecturers to summarize video lectures and assignments, enhancing learning efficiency. In summary, the future scope of the Video Summarization System revolves around enhancing models, expanding applications, streamlining data management, integrating into real-world scenarios, developing user-friendly interfaces, and fostering collaboration through open-sourcing. These advancements elevate the system's capabilities and ensure its adaptability to diverse user requirements in various industries.

REFERENCES

[1] V. Tiwari, and C. Bhatnagar, "A survey of recent work on video summarization: approaches and techniques", *Multimedia Tools Appl.*, vol. 80, no. 18, pp. 27187-27221, 2021. [http://dx.doi.org/10.1007/s11042-021-10977-y]

[2] E. Apostolidis, E. Adamantidou, A.I. Metsai, V. Mezaris, and I. Patras, "Video summarization using deep neural networks: A survey", *Proc. IEEE,* vol. 109, no. 11, pp. 1838-1863, 2021. [http://dx.doi.org/10.1109/JPROC.2021.3117472]

[3] D. Gupta, and A. Sharma, "A comprehensive study of automatic video summarization techniques", *Artif. Intell. Rev.*, vol. 56, no. 10, pp. 11473-11633, 2023. [http://dx.doi.org/10.1007/s10462-023-10429-z] [PMID: 36686596]

[4] J. Smith, and A. Johnson, "Video Summarization techniques: A comprehensive review", *Journal of Multimedia Processing,* vol. 7, no. 2, pp. 123-140, 1995.

[5] M. Anderson, R. Brown, and C. Davis, "Advancements in video processing: From keyframes to content-based retrieval", *Proceedings of the International Conference on Computer Vision,* 2000pp. 45-52.

[6] H. Thompson, and S. White, "Soccer highlights summarization using visual saliency", *Journal of Sports Technology,* vol. 12, no. 4, pp. 321-335, 2004.

[7] X. Chen, and Y. Kim, "Story-based video summarization: a hierarchical approach", *IEEE Trans. Multimed.,* vol. 8, no. 3, pp. 189-202, 2006.

[8] L. Garcia, and N. Patel, "Deep learning for image and video description: a survey", *Comput. Vis. Image Underst.,* vol. 131, pp. 32-49, 2014.

[9] Y. Liu, and M. Lapata, "BertSum: utilizing pretrained encoders for abstractive text summarization", *ACL Anthology,* pp. 567-578, 2018.

[10] Z. Wang, Q. Zhao, and H. Li, "Vision meets text: unsupervised video summarization with multimodal embeddings", *Int. J. Comput. Vis.,* vol. 112, no. 5, pp. 456-473, 2020.

[11] T. H. Brown, "GPT-3: language models are few-shot learners", *Proc Natl Acad Sci,* vol. 118, no. 9, p. e2020084118, 2021.

[12] Y. Liu, and M. Lapata, "Text summarization with pre-trained encoders", *arXiv,* p. 1908.08345, .

[13] S. Kumar, and A. Solanki, "An abstractive text summarization technique using transformer model with self-attention mechanism", *Neural Comput. Appl.,* vol. 35, no. 25, pp. 18603-18622, 2023. [http://dx.doi.org/10.1007/s00521-023-08687-7]

[14] T.H.M. Alcantara, D. Krütli, R. Ravada, and T. Hanne, "Multilingual Text Summarization for German Texts Using Transformer Models", *Information (Basel),* vol. 14, no. 6, p. 303, 2023. [http://dx.doi.org/10.3390/info14060303]

[15] S. Abdel-Salam, and A. Rafea, "Performance study on extractive text summarization using BERT models", *Information (Basel),* vol. 13, no. 2, p. 67, 2022. [http://dx.doi.org/10.3390/info13020067]

[16] Y. Jaafar, and K. Bouzoubaa, "Towards a new hybrid approach for abstractive summarization", *Procedia Comput. Sci.,* vol. 142, pp. 286-293, 2018. [http://dx.doi.org/10.1016/j.procs.2018.10.496]

[17] R. Rani, and S. Tandon, *Chat summarization and sentiment analysis techniques in data mining.,* 2018. [http://dx.doi.org/10.1109/ICCS.2018.00025]

[18] Y. Chen, Y. Liu, L. Chen, and Y. Zhang, "DialogSum: A real-life scenario dialogue summarization dataset," in Findings of the Association for Computational Linguistics: ACL-IJCNLP 2021, C. Zong, F. Xia, W. Li, and R. Navigli, Eds. Online: Association for Computational Linguistics, Aug. 2021, pp. 5062–5074. [http://dx.doi.org/10.18653/v1/2021.findings-acl.449]

[19] A. Sood, *Towards summarization of written text conversations.* International Institute of Information Technology: India, 2013.

[20] R. Goyal, P. Kumar, and V. P. Singh, "A systematic survey on automated text generation tools and techniques: Application, evaluation, and challenges", *Multimed Tools Appl,* vol. 82, no. 28, pp. 43089-43144, 2023. [http://dx.doi.org/10.1007/s11042-023-15224-0]

[21] N. Giarelis, C. Mastrokostas, and N. Karacapilidis, "Abstractive *vs.* extractive summarization: An experimental review", *Appl Sci,* vol. 13, no. 13, 2023.

[22] K.J. Han, R. Prieto, and T. Ma, "State-of-the-art speech recognition using multi-stream self-attention with dilated 1d convolutions", *2019 IEEE Automatic Speech Recognition and Understanding Workshop (ASRU),* IEEE., pp. 54-61, 2019. [http://dx.doi.org/10.1109/ASRU46091.2019.9003730]

CHAPTER 13

HDFS: The Backbone of Big Data - A Review of High Availability, Scalability, and Performance Using Quorum Journal Manager and Dynamic Federated Metadata Management

Bhagavathi Santos Kumar[1] and **Virendra Kumar Shrivastava**[1,*]

[1] *Department of Computer Science and Engineering, School of Advanced Computing, Alliance University, Bangalore, Karnataka 562106, India*

Abstract: Working in tandem, the Hadoop Distributed File System (HDFS) plays a crucial role in managing large-scale data across distributed clusters. The basic structure of HDFS comprises the NameNode, which manages the file system namespace and client interactions, and the Data Nodes, which handle storage and data arrangement duties. Although HDFS faces constraints, particularly regarding scaling the NameNode, which is responsible for managing metadata, these constraints impact namespace scalability and performance. Therefore, this paper discusses the two architectures, the Quorum Journal Manager (QJM) components and the active and standby NameNodes, which enable the sharing of edit logs between them. To mitigate the risk of metadata corruption from split-brain scenarios, the system ensures that only a single NameNode writes to the JournalNodes. Through the utilization of the QJM, HDFS attains High Availability (HA), enabling swift failover in the event of machine crashes or scheduled maintenance. Dynamic Federated Metadata Management (DFMM) addresses the limitations of the current Hadoop architecture by distributing metadata management across multiple federated components. It entails dispersing metadata management among various federated components. HDFS can overcome the problems of connected block storage, namespace scalability, and performance bottlenecks by dynamically managing metadata on an exabyte scale. This work summarizes the key findings and presents a comparative analysis of both architectures. QJM performs a vital function that ensures the high availability, reliability, and fault tolerance of the NameNode in Hadoop's HDFS. DFMM enhances the fault tolerance, scalability, and performance of distributed file systems by distributing metadata across multiple servers or nodes.

Keywords: DFMM, Dynamic federated metadata management, Hadoop, HDFS, HDFS federation, QJM, Quorum journal manager.

[*] **Corresponding Author Virendra Kumar Shrivastava:** Department of Computer Science and Engineering, School of Advanced Computing, Alliance University, Bangalore, Karnataka-562106, India; E-mail: Virendra.shrivastava@alliance.edu.in

Suneeta Satpathy, Sachi Nandan Mohanty & Subhendu Kumar Pani (Eds.)

INTRODUCTION

Hadoop and HDFS play a crucial role in managing tasks, providing expandable, resilient storage and processing capabilities [1] across distributed computer clusters. Hadoop is an open-source framework that facilitates the distributed handling of large datasets over computer clusters using a simplified programming model. The fundamental structure consists of Hadoop, which comprises HDFS [2] for storage, MapReduce for concurrent processing, and Yet Another Resource Negotiator (YARN) for resource management, facilitating adaptable and robust data management. The architectural design of the HDFS [3] is structured with an enslaver/agent configuration, wherein the NameNode functions as the primary server, managing the file system's namespace and client file access. The worker nodes, known as DataNodes, manage storage on the physical devices they control and divide data into blocks to enhance processing efficiency. A Secondary Name Node [4] is included in the design to assist with administrative tasks for the Name Node, including checking file system information at checkpoints. By enhancing HDFS's fault tolerance, this setup ensures that data remains available throughout the entire hardware cluster. The inability of HDFS to scale the NameNode, which is responsible for managing metadata [5], in the same manner as it scales the DataNodes is a drawback. The inability of the current Hadoop architecture to scale Name Nodes highlights the importance and challenges of managing metadata at the Exabyte scale. Consequently, this leads to closely interconnected block storage, constraints on namespace scalability, and performance [6] bottlenecks. Several limitations constrain the NameNode within HDFS. A single point of vulnerability results in system downtime when it fails or requires maintenance. Memory limitations restrict its capacity to accommodate files, particularly impacting performance when dealing with small files that may create bottlenecks. The centralized structure hinders scalability, while HDFS lacks the capabilities for low-latency data access, which is essential for real-time applications. Thus, architectures are available to mitigate the limitations of the NameNode, which are discussed in sections **3** and **4** below.

The subsequent parts of this paper are organized as follows. Section **2** outlines the main points of QJM and DFMM, discussing the importance of these systems in managing data and ensuring high availability. Section **3** provides a brief overview of QJM, and Section **4** discusses the HDFS Federation. Section **5** provides a brief overview of the DFMM and discusses the two main metadata management techniques in brief. Section **6** presents a contrast between QJM and DFMM. Section 7 discusses the implications of the findings. Section **8** describes the conclusion and future research prospects.

BACKGROUND

Table 1 presents a summary of potential related research papers. QJM plays a pivotal core component across the HDFS framework developed by Apache Hadoop. It aims to secure HA for the HDFS NameNode, addressing the SPOF challenge in standard HDFS setups. QJM achieves this by replicating the NameNode's edit log, which contains metadata changes, within a configurable quorum of nodes across the cluster. If the Active NameNode [13] becomes unavailable or is undergoing maintenance, a different node in the quorum will automatically assume control, maintaining uninterrupted access to the file system. DFMM [14] was developed to enhance the scalability [15] and performance of expensive distributed file systems, specifically when traditional metadata methods become bottlenecks. DFMM [16] distributes metadata and partitions among several metadata servers or nodes in a federated system. DFMM enhances the efficiency of metadata access and modification by distributing the workload across multiple servers, thereby improving scalability constraints associated with a centralized metadata server. This approach balances workloads, strengthens resilience against failures, and allows for seamless adjustment to evolving workload patterns and system conditions. These systems regularly integrate techniques. These systems regularly incorporate optimization techniques to ensure peak performance in dynamic computing environments. QJM and DFMM are vital for managing data and maintaining HA in distributed file systems such as HDFS.

Table 1. A summary of similar work that has been reviewed.

Reference	Approach	Description
[7]	High Availability	Proposed a hierarchical architecture of Hadoop Nodes for HA HDFS, eliminating single-point node failure and establishing a model for optimal data processing and storage.
[8]	Chord protocol-based Architecture	Reduces the reliance on the metadata volume on the NameNode and provides a highly scalable and fault-tolerant HDFS. The Chord protocol directly connects to the HDFS to address scalability issues. However, it also increases the complexity of the Single HDFS NameNode Architecture.
[9]	Quorum Journal Manager	Manages Journal Nodes carrying edit logs of the NameNode, mitigating the SPOF.
[10]	Triple H	A Hybrid Design (Triple H) for HDFS on the HPC cluster was proposed to enhance HDFS's efficiency.
[11]	HDFS Federation	This proposed architecture mitigates SPOF.
[12]	Dynamic Federated Metadata Management (DFMM)	This proposed architecture mitigates all the limitations of NameNode.

QUORUM JOURNAL MANAGER

In the HDFS framework, the primary objective of QJM accents is to provide high accessibility for the Name Node component. Inside the HDFS cluster, QJM makes sure of this by making a photocopy of the NameNode's edit log that involves metadata changes in the middle of a personalized kind of node. The edit log actively records all changes made to the file system, including activities such as creating, deleting, and modifying files. QJM ensures that every edit executed in the active NameNode is recorded in a separate storage area available to all nodes within the quorum. The divided storage solution can be implemented using a distributed file system, such as HDFS, or a Network Attached Storage (NAS) setup. The duplication of edit log entries to the quorum nodes can occur synchronously or asynchronously, depending on the configuration settings. If a failure or scheduled maintenance affects the active NameNode, another node within the quorum can assume the role of the active NameNode by retrieving the latest edit log segment from the communal storage. This seamless failover procedure ensures continuous access to the file system, eliminating any downtime periods.

The quorum-based storage design depends on achieving quorum agreement among a group of daemons known as JournalNodes. Every journal node provides a straightforward RPC interface enabling a NameNode to access and modify edit logs stored on the node's local disk. Upon editing, the NameNode dispatches the edit to every Journal Node within the cluster, then awaits acknowledgment from a majority (quorum) of the nodes. The edit can be deemed committed once most nodes confirm with a success code. The Quorum Journal Manager provides connections between the NameNode and Journal Nodes through RPC. This element integrates the pre-existing adaptable Journal Manager interface found in HDFS. It relies on implementing the File Journal Manager to handle its local storage. Fig. (**1**) presents the design of Quorum-based Storage.

Journal Node is a lightweight, standalone daemon that operates on an odd number of nodes. It offers actual storage of edit logs on the local disk. No additional hardware is required. The nodes operate on the same physical hardware as the NameNode, Standby NameNode, and JobTracker. To prevent split brain scenarios, where both NameNodes mistakenly assume the active role and attempt to write to the same file, which can lead to data corruption, JournalNode fencing using epoch numbers is implemented. Fig. (**2**) illustrates the design of the Quorum Journal Manager (QJM).

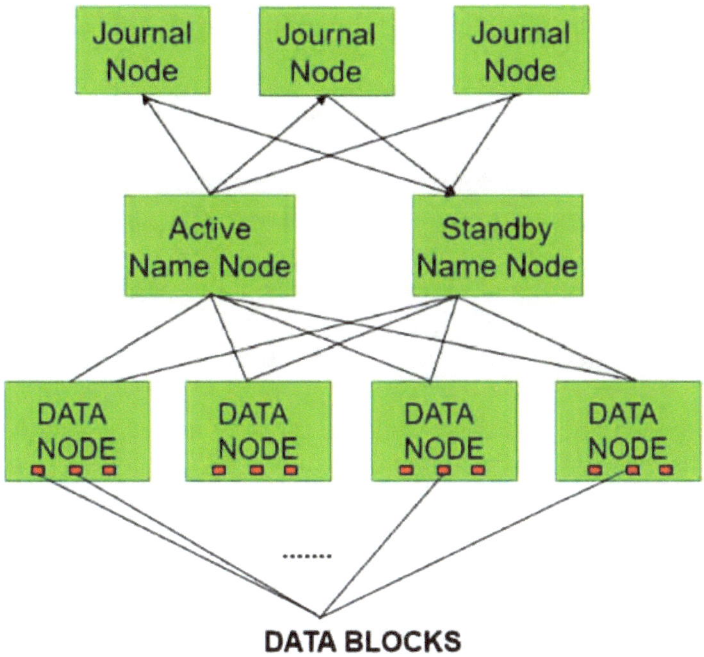

Fig. (1). Quorum-based storage.

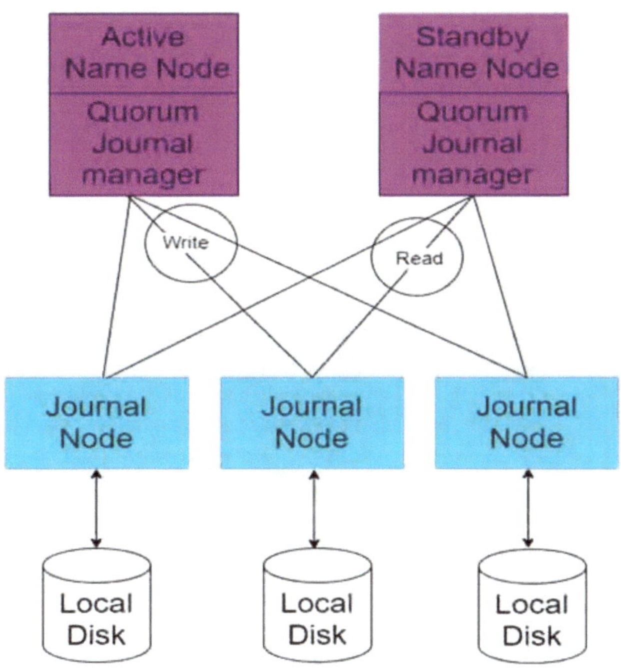

Fig. (2). Design of QJM.

The Design of QJM typically comprises the following fundamental components:

1. **Active NameNode:** The primary NameNode handles client requests and supervises the file system's metadata.
2. **Standby NameNodes:** Backup NameNodes maintain synchronization with the active NameNode by continuously duplicating the edit log entries from the quorum nodes.
3. **Quorum Nodes:** A group of nodes is responsible for storing and replicating the edit log segments. Together, these nodes form a quorum, and for an edit log entry to be committed, it must receive acknowledgment from a majority of the nodes.
4. **Shared Storage:** The nodes forming the quorum house and access the storage facility. This ability can be achieved using a distributed file system such as HDFS or NAS.

HA is guaranteed *via* QJM by facilitating a smooth transition to standby NameNodes and active NameNodes that experience failure or undergo planned maintenance, thus ensuring continuous accessibility to the file system. QJM enhances data integrity by duplicating the edit log across multiple quorum nodes, thereby decreasing the risk of data loss or corruption. Fault tolerance relies on quorum-based replication, where the majority of nodes must agree on the state of the edit log before it is considered committed. In the event of a failure, this mechanism ensures system robustness. The limitations of QJM are outlined in Table **2**.

Table 2. QJM potential limitations.

Limitations	Description
Resource Overhead	The development of this system may be attributed to the operation of several backup NameNodes and the duplication of log entries, resulting in increased demands on the cluster, including storage, capacity, and network speed.
Complexity	Handling QJM involves challenges that may require additional skills and resources, particularly in large-scale deployments characterized by complex network setups.
Synchronous Replication Overhead	The synchronous replication of edit log entries might impact the system performance by potentially causing delays in write operations.

HDFS FEDERATION

HDFS federation expands the capabilities of HDFS by dividing the file system namespace into distinct partitions, with each partition managed by its own NameNode. In the standard HDFS arrangement, a single NameNode oversees a solitary namespace. As the cluster size expands and the data volume increases, this setup poses a scalability bottleneck.

Key Features:

1. **Multiple Independent Namespaces:** HDFS federation expedites the creation of numerous independent namespaces within a unified HDFS cluster. Each namespace operates as a separate file system within the HDFS federation, featuring its own distinct directory structure, files, and metadata.
2. **Improved Scalability:** The limitation of traditional HDFS is mitigated by allowing for a divided namespace. Data and its management (metadata) are distributed across multiple NameNodes. This design fosters scalability, making the file system suitable for even massive datasets.
3. **Improved Fault Tolerance:** Consider a scenario where a NameNode fails. In a traditional HDFS setup, such a failure could disrupt the entire system. HDFS federation introduces fault isolation; with separate namespaces, a failure in one does not impact the others. This architecture ensures the availability and integrity of data in unaffected namespaces.
4. **Optimized Resource Usage:** HDFS federation tackles a bottleneck by distributing the metadata processing and storage workload. This eliminates single points of contention where resources get overloaded. By spreading the work across multiple NameNodes, the cluster's resources are used more efficiently.
5. **Compatibility:** HDFS Federation is engineered to maintain backward compatibility with current Hadoop applications and ecosystems. Without necessitating modifications, existing applications can benefit from these performance advantages.

The design of the HDFS federation incorporates numerous elements working together to provide a scalable and fault-tolerant distributed file system. (Fig. **3**) illustrates the design of the HDFS federation. The essential components are:

1. **NameNode:** In HDFS Federation, each namespace is overseen by a distinct NameNode instance. The NameNode is responsible for managing the metadata of files and directories within its associated namespace. Metadata encompasses file attributes (such as permissions and ownership), block locations, and namespace hierarchy. NameNodes integrate namespace operations, including file creation, deletion, renaming, and permission alterations. NameNodes conduct periodic checkpoints of their namespace state to assure durability and recoverability in the event of failure.
2. **DataNode:** Data Nodes are tasked with storing and serving the data blocks of files within the HDFS cluster. Every DataNode oversees a set of data blocks

and interacts with the relevant NameNodes to report block status and execute block operations. DataNodes handle client requests for reading and writing data, and they replicate blocks to ensure fault tolerance and data durability. Additionally, DataNodes perform block scanning and reporting to detect and address data corruption or failures.

3. **Namespace:** Within HDFS Federation, the entire file system is divided into multiple independent namespaces, each managed by a distinct NameNode. Every namespace functions as an independent file system within the HDFS Federation, processing its distinct directory structure, files, and metadata. Partitioning the Namespace enhances scalability and facilitates the isolation of data and metadata within the Hadoop cluster.

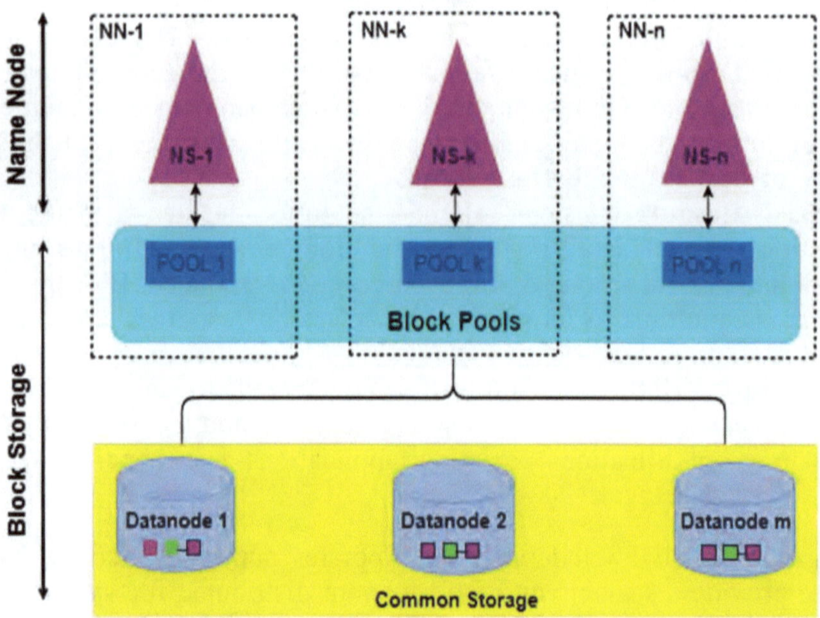

Fig. (3). Design of HDFS federation.

DYNAMIC FEDERATED METADATA MANAGEMENT

DFMM mitigates the scalability restrictions of centralized metadata management systems in distributed file systems. This technique improves scalability, performance, and fault tolerance. It mainly focuses on dynamically partitioning and propagating metadata across multiple servers or nodes. DFMM dynamically divides the file system's metadata and distributes it across numerous metadata servers or nodes within a federated architecture. Each metadata server handles a specific portion of the metadata file systems. The process can reduce the overall processing burden on a single server by distributing metadata operations across multiple servers. DFMM systems deal with the mechanisms for metadata synchronization and coordination to maintain consistency and coordination among the numerous metadata servers. These mechanisms employ distributed locking protocols, distributed transactions, or consensus algorithms, ensuring that metadata updates are uniformly and consistently applied across the metadata servers. Fig. (**4**) illustrates the design of DFMM.

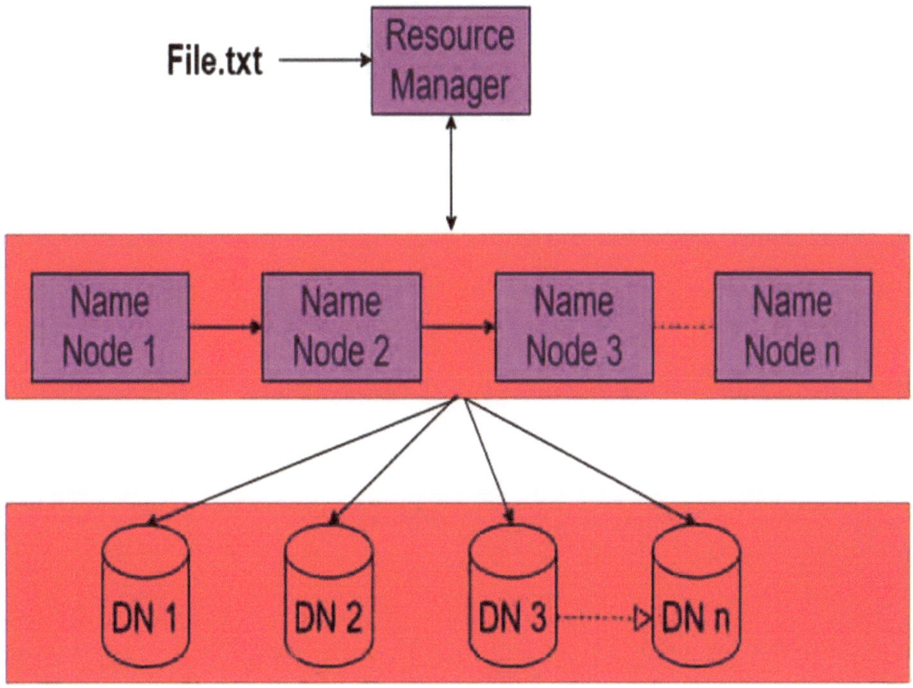

Fig. (4). Design of DFMM.

The advantages of DFMM are:

1. **Scalability:** DFMM increases the scalability of distributed file systems by separating the metadata management over various servers or nodes, allowing the file system to assist with increased metadata volumes and metadata-intensive operations.
2. **Performance:** By dividing metadata tasks over various servers, DFMM reduces the load on individual servers, amplifying the overall performance of the system and decreasing access latency.
3. **Fault Tolerance:** The alliance structure of DFMM upgrades the fault resistance by dividing metadata management tasks surrounded by various servers. If a server breaks, other servers can still handle metadata requests, guaranteeing continuous access to the file system. The limitations of the DFMM summary are presented in Table **3**.

3. DFMM potential limitations.

Limitations	Description
Complexity	Deploying and over-swinging an alliance metadata management system, such as DFMM, can be complicated, requiring conscientious design and cooperation over metadata servers.
Consistency Overhead	Ensuring consistency and coordination across metadata servers may entail additional communication and coordination overhead, potentially impacting system performance.
Synchronization Overhead	The synchronization mechanisms for metadata employed in DFMM systems could potentially cause delays in metadata operations, especially in situations with prominent concurrency levels or frequent metadata updates.

Metadata Management Techniques

There are two efficient management techniques for managing distributed metadata [17 - 21], namely, sub-tree partitioning and Hashing techniques.

Subtree Partitioning Technique

Using the sub-tree partitioning technique [21] [22], hashing directories on the storage server partition metadata. If a particular server experiences a higher data load, the central server automatically relocates all the subdirectories to an available server.

This technique distributes load evenly among servers when workloads are evenly spread. The RAM on each server is utilized efficiently through a prefix caching mechanism. However, the main disadvantage of this technique is that the workload is distributed unevenly among metadata servers, resulting in server bottleneck issues. For example, in Fig. (**5**), the namespace partition is shown according to each server's directory subtree, and the directory subtree is hashed on other metadata servers if any servers are under heavy load. Each metadata server (MDS) manages the metadata for an entire directory tree. However, if an application repeatedly visits the file, directory, or subtree, it can lead to server bottleneck issues.

Hashing Technique (HT)

In the Hashing Technique [23], the file's metadata is stored in a specific metadata server, which uses the hash function on the file. Lustre and ZFS utilize a hash function to locate metadata using path name hashing. Compared to sub-tree partitioning, the Hashing Technique enables uniform server distribution. The disadvantage of Hashing is that the hash function is randomly assigned; hence, while performing metadata operations, it leads to network overhead. The hashing process is shown in Fig. (**6**).

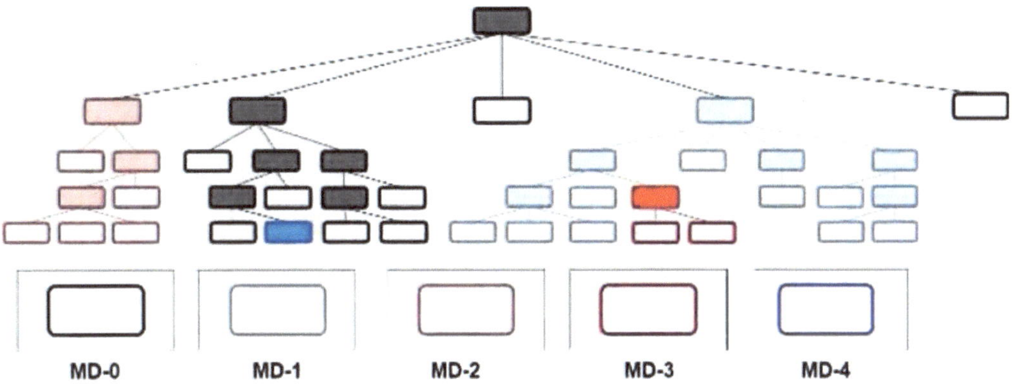

Fig. (5). Sub-tree partitioning.

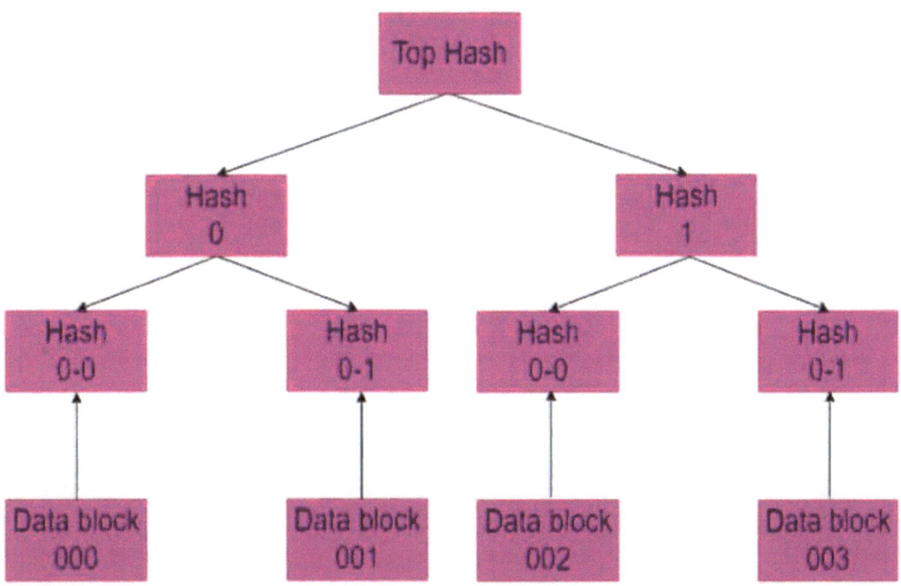

Fig. (6). Hashing process.

The hashing technique uses a hash value to partition the namespace among the various metadata servers. However, it is necessary to maintain the directory hierarchy, semantics, and access permissions. Let the hash value be the same as the file name. In this case, the duplicate files will be allocated to different metadata servers, even though the file belongs to other directories. This leads to a server bottleneck problem. To proficiently manage the data and metadata of distributed storage systems and improve their scalability and performance, the sharding approach is combined with Hashing and Sub-Tree partitioning techniques.

CONTRAST BETWEEN QUORUM JOURNAL MANAGER (QJM) AND DYNAMIC FEDERATED METADATA MANAGEMENT (DFMM)

The contrast between QJM and DFMM is presented in Table **4**.

Table 4. Contrast between QJM and DFMM.

Feature	Quorum Journal Manager (QJM)	Dynamic Federated Metadata Management (DFMM)
Purpose	Assurance of the Continuous availability of NameNode in HDFS	Enhance the scalability and performance of metadata management.
Data Management	Duplicate the NameNode's edit log to ensure fault tolerance	Dynamically divides and disperses metadata among various servers
High Availability	Facilitates a smooth transition to standby NameNodes in the event of failure.	Improves fault tolerance through the distribution of metadata management tasks
Architecture	Consolidated with both active and standby NameNodes	Decentralized with numerous metadata servers managing subsets of metadata
Key Components	Primary NameNode, Backup NameNodes, Quorum Servers, Shared Storage	Servers managing metadata, Partitioning mechanisms, Co-ordination mechanisms, Caches for Metadata
Advantages	Continuous availability, data consistency, and resilience.	Expandability, efficiency, resilience
Potential Limitations	Resource utilization, intricacy, and latency of synchronous replication.	Complexity, coordination overhead, synchronization delay

The primary purpose of QJM and DFMM is to improve the accuracy and scalability of distributed file systems. They address fundamental requirements in distributed file systems, including high availability, fault tolerance, and scalability. Both systems involve repeated mechanisms to ensure data integrity and fault tolerance, though they relate to separate elements of the file system (edit log for QJM and metadata for DFMM). QJM focuses on replicating the NameNode's edit log to ensure high availability for the NameNode. In contrast, DFMM concentrates on distributing metadata management tasks across multiple servers to enhance scalability and performance. QJM relies on a quorum-based replication mechanism and backup NameNodes to ensure high availability, whereas DFMM utilizes a federated architecture with numerous metadata servers. QJM primarily addresses the single point of failure issue associated with the NameNode, whereas DFMM focuses on overcoming the scalability limitations of centralized metadata management systems.

IMPLICATIONS AND FINDINGS

By copying the edit log from the NameNode, QJM initially focuses on ensuring that the Hadoop-distributed file system remains accessible. It improves fault tolerance and identifies every point of failure. QJM data management technique is based on replicating changes to the metadata across most nodes, ensuring data stability and integrity during maintenance or failures. By dynamically splitting and distributing metadata across multiple servers or nodes [24 - 26], DFMM aims to enhance the scalability and efficiency of metadata management. The goal of DFMM's approach is to enhance performance by reducing the workload on each server, handling metadata individually, and distributing the effort across multiple servers.

Specific Use Cases and Preferences

Quorum Journal Manager (QJM)

- High accessibility and fault tolerance are ideal in scenarios where they are most important, such as mission-critical programs or settings that cannot tolerate downtime.
- Compatible with deployments prioritizing continuous file system access and safeguarding integrity in the event of planned maintenance or failures.

Dynamic Federated Metadata Management (DFMM)

- Ideal for environments where traditional centralized information management techniques create bottlenecks, such as those with high frequency and extensive metadata requirements.
- Suitable for deployments in which distributed workload management and scalability performance increase are less important factors, such as cloud-based storage systems or multitenant surroundings with diverse workloads.

CONCLUSION AND FUTURE DIRECTION

Regarding data management and high accessibility, both QJM and DFMM are essential in addressing significant issues in distributed file systems. By replicating the Name Node's edit log across a quorum of nodes, QJM aims to ensure high availability for the Name Node component of Hadoop's HDFS, while enhancing fault tolerance and ensuring unimpeded access to the file system. DFMM divides and distributes metadata among multiple servers or nodes to improve the speed and scalability of distributed file systems. This technique enhances scalability and more effective metadata management in large-scale distributed storage setups. When choosing between QJM and DFMM, it is vital to consider data processing and system reliability when using distributed file systems.

QJM provides a robust solution for high availability and fault tolerance, particularly in Hadoop-based systems, ensuring consistent file system performance and access even during Name Node maintenance jobs or outages. On the other hand, companies managing large-scale distributed storage setups and facing scalability issues with central metadata management may benefit from implementing DFMM. Its scalability and speed are improved by distributing the burden of information management across several servers, which enables it to manage massive amounts of metadata and metadata-intensive operations. There are several key areas of study to investigate in this field, and the future appears promising.

Enhancements of QJM: Enhancing and optimizing QJM for large-scale deployments to improve performance, scalability, and resource efficiency. Developments in algorithms, data structures, or implementation strategies may be involved.

Advancements of DFMM: Creative partitioning and distribution techniques for DFMM are being developed to improve metadata management in distributed file systems. This study considers variables such as the nature of the workload system setups and user data access.

Effects of emerging technologies: Evaluate the impact of emerging technologies on the design and implementation of high availability and data management solutions for distributed file systems, including edge computing and containerization. Future system design may face new opportunities and difficulties as a result of this investigation.

REFERENCES

[1] I. Khatri, and V.K. Shrivastava, "A survey of big data in the healthcare industry", *Proceedings of the 9th ICACCT,* 2016pp. 245-257. [http://dx.doi.org/10.1007/978-981-10-1023-1_25]

[2] K. Shvachko, H. Kuang, S. Radia, and R. Chansler, "The Hadoop Distributed File System," 2010 IEEE 26th Symposium on Mass Storage Systems and Technologies (MSST), Incline Village, NV, USA, 2010, pp. 1-10. [http://dx.doi.org/10.1109/MSST.2010.5496972]

[3] Vishal Gaurav, *Offline context aware computing in smart devices for intelligent reminders.,* 2015. [http://dx.doi.org/10.1109/ICECCT.2015.7226118]

[4] P.M. Dhulavvagol, V.H. Bhajantri, and S.G. Totad, "Blockchain ethereum clients performance analysis considering e-voting application", *Procedia Comput. Sci.,* vol. 167, pp. 2506-2515, 2020. [http://dx.doi.org/10.1016/j.procs.2020.03.303]

[5] F. Ullah, and M. A. Babar, *An architecture-driven adaptation approach for big data cyber security analytics.,* 2019. [http://dx.doi.org/10.1109/ICSA.2019.00013]

[6] Abhinandan Ramaprasath, *Performance optimization of big data in mobile networks.,* 2015. [http://dx.doi.org/10.1109/CCECE.2015.7129477]

[7] S. Chakraborty, K. Barua, M. Pandey, and S. Rautaray, "A Proposal for High Availability of HDFS Architecture based on Threshold Limit and Saturation Limit of the Namenode", *International Journal of Information Engineering and Electronic Business,* vol. 9, no. 6, pp. 27-34, 2017. [http://dx.doi.org/10.5815/ijieeb.2017.06.04]

[8] Z. Wang, and D. Wang, *NCluster: Using Multiple Active Name Nodes to Achieve High Availability for HDFS.,* 2013. [http://dx.doi.org/10.1109/HPCC.and.EUC.2013.329]

[9] Y. Kim, "A Distributed NameNode Cluster for a Highly-Available Hadoop Distributed File System", In: *2014 IEEE 33rd International Symposium on Reliable Distributed Systems (SRDS).* IEEE, 2014. [http://dx.doi.org/10.1109/SRDS.2014.61]

[10] N. S. Islam, X. Lu, M. Wasi-ur-Rahman, D. Shankar and D. K. Panda, "Triple-H: A Hybrid Approach to Accelerate HDFS on HPC Clusters with Heterogeneous Storage Architecture," 2015 *15th IEEE/ACM International Symposium on Cluster, Cloud and Grid Computing,* Shenzhen, China, 2015, pp. 101-110.

[11] P.M. Dhulavvagol, S.G. Totad, A.S. Meti, and V. Shashidhara, "An Adaptive and Dynamic Dimensionality Reduction Method for Efficient Retrieval of Videos", *Commun. Comput. Inf. Sci.,* vol. 1035, pp. 624-639, 2019. [http://dx.doi.org/10.1007/978-981-13-9181-1_54]

[12] I. Polato, R. Ré, A. Goldman, and F. Kon, "A comprehensive view of Hadoop research—A systematic literature review", *J. Netw. Comput. Appl.,* vol. 46, pp. 1-25, 2014. [http://dx.doi.org/10.1016/j.jnca.2014.07.022]

[13] X. Wang, and J. Su, *Research of Distributed Data Store Based on HDFS.,* 2013. [http://dx.doi.org/10.1109/ICCIS.2013.384]

[14] P.M. Dhulavvagol, and S.G. Totad, "Performance enhancement of distributed system using hdfs federation and sharding", *Procedia Comput. Sci.,* vol. 218, pp. 2830-2841, 2023. [http://dx.doi.org/10.1016/j.procs.2023.01.254]

[15] D. Stamatakis, N. Tsikoudis, O. Smyrnaki, and K. Magoutis, "Scalability of replicated metadata services in distributed file systems," in *Distributed Applications and Interoperable Systems (DAIS 2012),* K. M. Göschka and S. Haridi, Eds., Lecture Notes in Computer Science, vol. 7272. Berlin, Heidelberg: Springer, 2012. [http://dx.doi.org/10.1007/978-3-642-30823-9_3]

[16] A. D. Popescu, A. Balmin, V. Ercegovac, and A. Ailamaki, "PREDIcT: Towards predicting the runtime of large scale iterative analytics," *Proceedings of the VLDB Endowment,* vol. 6, no. 14, pp. 1678–1689, Sept. 2013. [http://dx.doi.org/10.14778/2556549.2556553]

[17] D. Stamatakis, N. Tsikoudis, E. Micheli, and K. Magoutis, "A General-Purpose Architecture for Replicated Metadata Services in Distributed File Systems", *IEEE Trans. Parallel Distrib. Syst.,* vol. 28, no. 10, pp. 2747-2759, 2017. [http://dx.doi.org/10.1109/TPDS.2017.2700272]

[18] S. A. Brandt, E. L. Miller, D. D. E. Long, and Lan Xue, *Efficient metadata management in large,*

distributed storage systems., 2003.
[http://dx.doi.org/10.1109/MASS.2003.1194865]

[19] M. Ismail, S. Niazi, G. Berthou, M. Ronström, S. Haridi, and J. Dowling, "HopsFS-S3: Extending Object Stores with POSIX-like Semantics and more (industry track)", *Proceedings of the 21st International Middleware Conference Industrial Track,* 2020pp. 23-30
[http://dx.doi.org/10.1145/3429357.3430521]

[20] P. Macko, and J. Hennessey, "Survey of distributed file system design choices", *ACM Trans. Storage,* vol. 18, no. 1, pp. 1-34, 2022.
[http://dx.doi.org/10.1145/3465405]

[21] A. Vattani, "k-means requires exponentially many iterations even in the plane", *Discrete Comput. Geom.,* vol. 45, no. 4, pp. 596-616, 2011.
[http://dx.doi.org/10.1007/s00454-011-9340-1]

[22] P.M. Dhulavvagol, and S.G. Totad, "Topic based partitioning for selective search using sharding technique", *2022 International Conference for Advancement in Technology (ICONAT),* 2022
[http://dx.doi.org/10.1109/ICONAT53423.2022.9726020]

[23] D. Karger, E. Lehman, T. Leighton, R. Panigrahy, M. Levine, and D. Lewin, "Consistent hashing and random trees", *Proceedings of the Twenty-Ninth Annual ACM Symposium on Theory of Computing - STOC '97,* 1997
[http://dx.doi.org/10.1145/258533.258660]

[24] H. Dai, Y. Wang, K.B. Kent, L. Zeng, and C. Xu, "The state of the art of metadata managements in large-scale distributed file systems — scalability, performance and availability", *IEEE Trans. Parallel Distrib. Syst.,* vol. 33, no. 12, pp. 3850-3869, 2022.
[http://dx.doi.org/10.1109/TPDS.2022.3170574]

[25] P. Somasekaram, R. Calinescu, and R. Buyya, "High-availability clusters: A taxonomy, survey, and future directions", *J. Syst. Softw.,* vol. 187, p. 111208, 2022.
[http://dx.doi.org/10.1016/j.jss.2021.111208]

[26] P.S. K S, P.M. Javali, R. R C, R.S. Shet, and N. Madhu, "A literature review of big data and metadata management systems", *Int. J. Res. Appl. Sci. Eng. Technol.,* vol. 11, no. 3, pp. 252-261, 2023.
[http://dx.doi.org/10.22214/ijraset.2023.49294]

CHAPTER 14

Optimized Gold Price Prediction Using Particle Swarm Optimization-Enhanced LSTM Networks

Alladi Vennela[1] and **Radha Mohan Pattanayak**[1,*]

[1] *School of Computer Science and Engineering (SCOPE), VIT-AP University, Amaravati, Andhra Pradesh-522237, India*

Abstract: Gold has long been regarded as a valuable asset and a reliable haven for investors, particularly during periods of economic uncertainty. Gold price forecasting is difficult due to the market's intrinsic volatility and sensitivity to various factors, including global economic trends, inflation rates, currency value fluctuations, and geopolitical events. This study presents a novel strategy for predicting gold prices that combines Long Short-Term Memory (LSTM) neural networks and Particle Swarm Optimization (PSO). Traditional models struggle with these complications of uncertainty, but the PSO-LSTM technique improves LSTM hyperparameters for greater accuracy. The PSO-LSTM model beats both standalone deep Learning models and traditional forecasting, based on performance criteria such as Mean Absolute Error (MAE), Root Mean Square Error (RMSE), and Mean Absolute Percentage Error (MAPE).

Keywords: Deep learning, Gold price, LSTM, Machine learning, Optimization, Time series.

INTRODUCTION

Gold has been a fundamental pillar in the intricate web of global economies and financial markets for a long time, symbolizing prosperity, firmness, and security. It is of great significance as one of the valuable metals utilized for financing trade transactions. In countries like the United States, India, the United Arab Emirates, China, and many others, it is highly esteemed for its role in jewelry. Furthermore, gold is commonly used as a valuable present or symbol of memory, and gold ornaments are frequently included in matrimonial agreements [1]. As more people become aware of and believe that gold can be owned as a low-risk investment asset, demand for gold is increasing daily. To mitigate financial losses, it is crucial to predict the value of gold [2] accurately.

* **Corresponding author Radha Mohan Pattanayak:** School of Computer Science and Engineering (SCOPE), VIT-AP University, Amaravati, Andhra Pradesh-522237, India; E-mail: radhamohan.pattanayak@gmail.com

Suneeta Satpathy, Sachi Nandan Mohanty & Subhendu Kumar Pani (Eds.)

Researchers are currently observing a growing interest in gold acquisition, focusing on comprehending the specific characteristics that influence price fluctuations. Previously, numerous approaches were studied to predict the future value of gold, including Machine Learning models [3] such as ARIMA [4, 5], Least Square Support Vector Machine [6], Tree based models [7, 8], Multi Linear Regression Model [9, 10], Multivariate Stochastic Model [11, 12], Extreme Learning Machine [13, 14] and Fuzzy set-based models have also been widely used for time series predictions, such as high-order fuzzy time series [15, 16], hesitant fuzzy time series [17, 18], and neutrosophic fuzzy models [19, 20], demonstrating their effectiveness in forecasting complex datasets.

The proposed PSO-LSTM model aims to assist investors in making informed decisions by accurately predicting future gold prices. This article explains how PSO is combined with LSTM, describes the dataset and implementation steps, and reviews the experimental setup and performance results. The findings are discussed, and suggestions for future research in gold price prediction are provided.

The paper is organized as follows: Section **2** discusses earlier gold price forecast methodologies. Section **3** describes the methods employed in this investigation. Section **4** contains the findings and discussions, which compare the proposed model to benchmarks. Finally, Section **5** presents the key findings and offers suggestions for future research topics.

LITERATURE REVIEW

Estimating the prices of gold is of great importance to investors, financial institutions, and policymakers, leading to extensive studies in this area. This section provides a comprehensive review of the existing literature on gold price forecasting, encompassing methodologies, strategies, and key findings from previous studies. Traditional econometric models, such as ARIMA [21], have been extensively used for forecasting gold prices. Tanattrin Bunnag utilized ARIMA models to forecast gold prices based on historical data, achieving reasonable accuracy in short-term predictions [22].

Wenjing Fang proposed three methods for predicting gold prices: regression analysis, back propagation neural network (BP neural network), and time series analysis—the gold price dataset is used for prediction with high reliability and accuracy [23]. Minal Ghute and Mridula S. Korde analyzed economic variables and gold prices using machine learning algorithms. It compares four algorithms —decision tree, linear regression, random forest, support vector machine, and ridge regression —to predict future gold prices accurately [24].

Using machine learning algorithms, D. Nanthiya, S. B. Gopal, S. Balakumar, M. Harisankar, and S. P. Midhun. developed a model to predict daily and future gold prices. The study evaluated the performance of Linear Regression, ARIMA, and Random Forest models. The gold price dataset was used to assess accuracy, yielding the following results: Linear Regression achieved an MAE of 19.82 and an RMSE of 24.41, the ARIMA model had an MAE of 0.040 and an RMSE of 0.046, and the Random Forest model recorded an MAE of 0.150 and an RMSE of 0.156 [25].

Dasari Siva Sankar and Harish Sahu suggested a Random Forest regressor approach for predicting gold prices. The R-squared error was utilized to study the influence of the number of estimators on accuracy and prediction time, and the Random Forest model was trained on a dataset with a relative squared error of 0.99077 [26]. Ziyang Yuan implemented KNN, XGBoost, and Light GBM models to predict Bitcoin and Gold prices. The OpenCV dataset was utilized for accuracy in predicting Bitcoin and gold prices. Results guide investors and offer ML insights for price forecasting [27].

METHODOLOGY

Data Collection

The dataset from Kaggle contains historical gold prices with key attributes: Date, Price (closing), Open, High, Low, Volume, and Change Percentage. The 'Date' column represents the trading day, while 'Price' shows the closing price. 'Open,' 'High,' and 'Low' reflect the session's opening, highest, and lowest prices. 'Volume' indicates the total gold traded, and 'Change %' represents the price change compared to the previous day. The data can be collected from Kaggle platforms that provide historical gold price records over a desired period. Fig. (**1**) below represents the flowchart of the proposed model.

Data Preparation

The data must be pre-processed before being fed into the prediction model to ensure its quality and usefulness.

- Data Cleaning: Handling missing values or outliers that may exist in the dataset.
- Normalization: Scaling the data within a specific range, commonly between 0 and 1, to help the LSTM model perform optimally.
- Feature Engineering: Extracting essential features from the raw data to improve the model's predictive power. For instance, additional features could be created based on moving averages, momentum indicators, or lagged price data.

- Data Splitting: The dataset will be split into two subsets, one for training the LSTM model and the other for testing its performance. The training subset is utilized to help the model learn patterns, while the testing subset assesses the model's predictive accuracy.

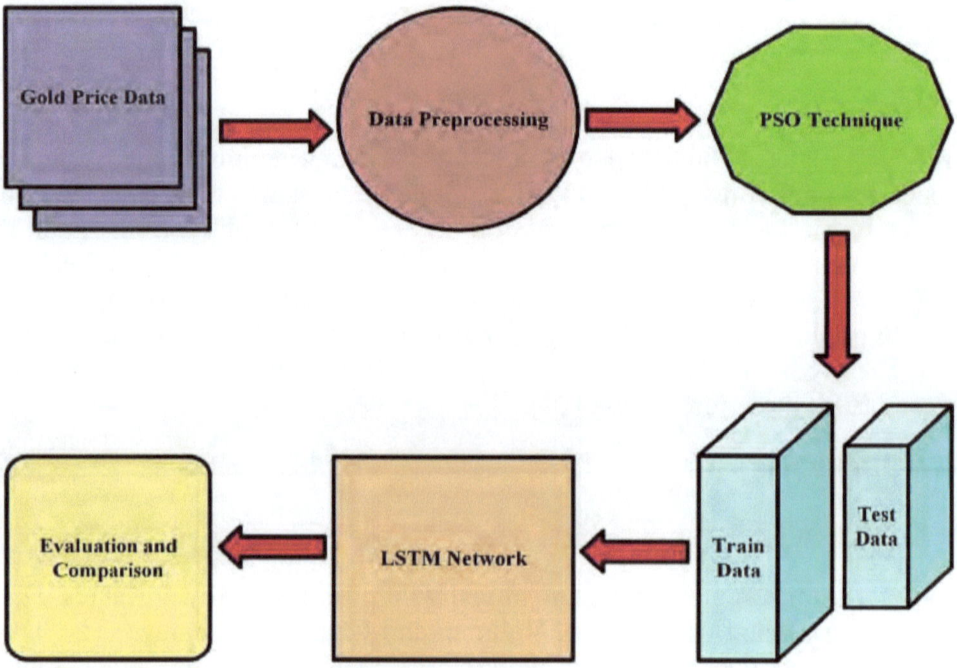

Fig. (1). Proposed PSO-LSTM model flowchart.

Particle Swarm Optimisation (PSO)

PSO is used to optimize key hyperparameters of the LSTM network, such as the number of layers, neurons per layer, learning rate, epochs, and batch size. By efficiently navigating the hyperparameter space, PSO improves the model's performance compared to traditional tuning methods. Each particle in the PSO algorithm represents a possible solution to the hyperparameter optimization problem. These particles traverse the search space, adjusting their velocities based on their past success and the best solutions discovered by other particles in the swarm.

As the particles move through this space, they continuously update their positions by considering their individual experiences and the group's collective knowledge, including the best solutions. This collaborative search enables the PSO algorithm to quickly converge on optimal hyperparameter settings that reduce the LSTM model's prediction error. Through this adaptive process, PSO accelerates the tuning phase and helps avoid local optima, leading to more accurate and generalizable predictions on unseen data.

Train Data & Test Data

- With the help of Train Data, the LSTM model is trained on a subset of pre-processed data to identify patterns and correlations.
- Separate from training, this subset evaluates the model's ability to generalize to new data. This helps assess the model's accuracy and reliability in real-world settings.

Long Short-Term Memory (LSTM) network

The LSTM network is a type of recurrent neural network (RNN) that excels at forecasting time series data, such as gold prices. The LSTM can learn and retain both short- and long-term dependencies in data, making it well-suited for complex prediction tasks, such as stock or gold price forecasting, with the help of the following layers:

- **Input Layer:** The input layer processes sequential data, such as time series or text sequences. The input is processed one step at a time. The LSTM network receives historical gold price data as input.
- **LSTM Cell:** The LSTM cell has four gates: Forget Gate, Input Gate, Cell Gate, and Output Gate. The Forget Gate discards information from the previous time step, the Input Gate controls how much new information should be added to the cell state, the Cell State stores essential information across time steps, and the Output Gate determines which part of the cell state should be output to the next time step and hidden layer.
- **Hidden Layer:** Stores the hidden states (short-term memory), which are updated at each time step. These hidden states are passed through the network to capture patterns over time.
- **Output Layer:** Produces the final output of the network, often a predicted value in forecasting tasks, such as a future price in financial data.

LSTM is chosen because it can handle the vanishing gradient problem that often affects traditional RNNs. It is well-suited for sequential data tasks, such as time series forecasting. The LSTM network predicts the future gold price based on the input data and optimized hyperparameters obtained from PSO.

Evaluation and Comparison

Evaluation: After training, the model's performance is tested against test data using measures like Mean Squared Error, Root Mean Squared Error, Mean Absolute Error, R-squared Error

Comparison: The performance of the PSO-enhanced LSTM model is compared with other models to demonstrate its effectiveness in predicting gold prices.

RESULTS AND DISCUSSIONS

Results

Analyzing the presented technique's performance demonstrates its advantage over earlier works. Various performance metrics are used to assess the effectiveness of the stated work, including MSE, MAE, MAPE, RMSE, and R^2. The suggested work is carried out using Python.

The implemented model achieved high accuracy and minimal loss over the dataset of gold price data (from 2013 to 2023). In both testing and training sets, the proposed model produced better performance. The proposed model achieved an accuracy of more than 98% and produced minimal losses, with a value of 0.02.

Performance Metrics

The suggested work has been assessed using the MSE (1), RMSE (2), MAPE (3), and R-square error metrics (4), which are presented below:

$$MSE = \frac{1}{n} \sum_{l=1}^{V} \left(X_a - \hat{X}_a\right)^2 \tag{1}$$

$$RMSE = \sqrt{\frac{1}{V} \sum_{l=1}^{V} \left(X_a - \hat{X}_a\right)^2} \tag{2}$$

$$MAPE = \frac{100}{n} \sum_{l=1}^{n} \frac{|X_a - \hat{X}_a|}{X_a} \tag{3}$$

Where:

$$MAPE = 1 - \frac{\Sigma \quad (X_a - \hat{X}_a)^2}{\Sigma \quad (X_a - \hat{X}_a)^2} \tag{4}$$

- X_a is represented by the original/actual prediction.
- X_a is represented the forecasted outcome.
- n represents the number of observations.

Discussion

The proposed PSO and LSTM model effectively predicted future gold prices by leveraging the strengths of PSO and LSTM networks. Traditional techniques often face high error rates, significant computational costs, and increased complexity. The PSO algorithm tackles these concerns by optimizing hyperparameters in the LSTM model, consequently improving the predictive performance and efficiency of the proposed model. The PSO-LSTM model performance has been evaluated with various metrics, demonstrating its superior accuracy in gold price prediction. The proposed model has achieved the lowest RMSE of 0.0191, MSE of 0.0075, and MAPE of 0.0061, along with a high R-squared (R^2) score of 0.9441 on the gold price dataset, demonstrating its strong predictive capacity for the variance in gold price data when compared with existing models. The results of the proposed method, along with a comparison with existing methods, are detailed in the Table **1**.

Table 1. Comparison of proposed work with existing methods.

<table>
<tr><th>S. No</th><th>Author</th><th>Techniques used</th><th>MSE value</th><th>RMSE value</th><th>R2 value</th><th>MAPE value</th></tr>
<tr><td rowspan="5">1</td><td rowspan="5">N. Tripurana, B. Kar, S. Chakravarty, B. K. Paikaray, and S. Satpathy</td><td>Random Forest regression</td><td>0.0065</td><td>0.0808</td><td>0.6049</td><td>1481</td></tr>
<tr><td>Decision Tree Regression</td><td>0.0021</td><td>0.0461</td><td>0.3399</td><td>0.0421</td></tr>
<tr><td>Support Vector Regression</td><td>0.0138</td><td>0.1177</td><td>0.1026</td><td>0.1153</td></tr>
<tr><td>Linear Regression</td><td>0.0002</td><td>0.0163</td><td>0.9169</td><td>0.0137</td></tr>
<tr><td>Artificial Neural Networks</td><td>0.0199</td><td>0.0269</td><td>0.7738</td><td>0.0217</td></tr>
<tr><td>2</td><td>D. Sarvaiya and D. Ramchandani [28]</td><td>Hybrid ARIMA and LSTM model</td><td>3239</td><td>5691</td><td>0.5776</td><td>0.0366</td></tr>
<tr><td>3</td><td>A. Kumar Agarwal, S. Kumari [29]</td><td>Hybrid Linear Regression and ARIMA model</td><td>7426</td><td>86.171</td><td>-1.59</td><td>0.0366</td></tr>
<tr><td>4</td><td>T. Tobin and R. Kashef [30]</td><td>A hybrid method of CNN and RNN</td><td>0.0092</td><td>0.0961</td><td>0.5422</td><td>0.0664</td></tr>
<tr><td>5</td><td>P. K. S. and L. R. S. Goel, M. Saxena [31]</td><td>RNN</td><td>0.0005</td><td>0.0238</td><td>0.0217</td><td>0.9530</td></tr>
<tr><td>6</td><td>Proposed Work</td><td>PSO-LSTM</td><td>0.0075</td><td>0.0191</td><td>0.9441</td><td>0.0061</td></tr>
</table>

CONCLUSION

In conclusion, predicting gold prices using Particle Swarm Optimization (PSO) and deep learning techniques offers a robust and practical approach. The integration of PSO helps optimize the DL model's hyperparameters, enhancing its efficacy. By leveraging the advantages of deep learning in understanding complex patterns in historical data and the optimization capabilities of PSO, this hybrid method can deliver more precise and reliable predictions. Therefore, it is a valuable tool for traders and economists to make informed choices in the unpredictable gold market. Future work could focus on enhancing the hybrid PSO-deep learning model for gold price prediction by integrating additional data sources, like macroeconomic indicators, Political disturbances, and financial perceptions derived from social media and news articles.

REFERENCES

[1] N. Tripurana, B. Kar, S. Chakravarty, B. K. Paikaray, and S. Satpathy, "Gold price prediction using machine learning techniques," in *CEUR Workshop Proceedings*, 2021, pp. 274–281. [http://dx.doi.org/10.54097/gdm0kc53]

[2] P.S.A. Surya Pangestu, A. Rochman, and A. Zuhdi, "The comparison of gold price prediction techniques using long short term memory (LSTM) and fuzzy time series (FTS) method", *Intelmatics,* vol. 3, no. 2, pp. 57-62, 2023. [http://dx.doi.org/10.25105/itm.v3i2.17325]

[3] Aruna S, Sujipriya J, and R. Umamaheswari, "Prediction of Potential Gold Prices using Machine Learning Approach," *Ann. Rom. Soc. Cell Biol.,* vol. 25, pp. 1385–1396, Apr. 2021, Accessed: Jul. 01, 2024. [Online]. Available from: http://www.annalsofrscb.ro/index.php/journal/article/view/2645

[4] R.M.P. Pattanayak, M.V. Sangameswar, D. Vodnala, and H. Das, "Fuzzy time series forecasting approach using LSTM model", *Comput. Sist.,* vol. 26, no. 1, pp. 485-492, 2022. [http://dx.doi.org/10.13053/cys-26-1-4192]

[5] G. Bandyopadhyay, "Gold Price Forecasting Using ARIMA Model", *Journal of Advanced Management Science,* vol. 4, no. 2, pp. 117-121, 2016. [http://dx.doi.org/10.12720/joams.4.2.117-121]

[6] S. Panigrahi, R.M. Pattanayak, P.K. Sethy, and S.K. Behera, "Forecasting of Sunspot Time Series Using a Hybridization of ARIMA, ETS and SVM Methods", *Sol. Phys.,* vol. 296, no. 1, p. 6, 2021. [http://dx.doi.org/10.1007/s11207-020-01757-2]

[7] G.S. Pillai, and M.I. Mary, *Intelligent systems and applications in engineering gold price forecast based on the least square support vector machine,* vol. 12, no. 3, pp. 684-697, 2024.

[8] R. M. Pattanayak, H. S. Behera, and S. Panigrahi, "A novel probabilistic intuitionistic fuzzy set based model for high order fuzzy time series forecasting," *Eng. Appl. Artif. Intell.,* vol. 99, no. December 2020, p. 104136, 2021, [http://dx.doi.org/10.1016/j.engappai.2020.104136]

[9] B.K. Padhi, S. Chakravarty, B. Naik, R.M. Pattanayak, and H. Das, "RHSOFS: Feature Selection Using the Rock Hyrax Swarm Optimization Algorithm for Credit Card Fraud Detection System", *Sensors (Basel),* vol. 22, no. 23, p. 9321, 2022. [http://dx.doi.org/10.3390/s22239321] [PMID: 36502020]

[10] P. Baser, J.R. Saini, and N. Baser, "Gold commodity price prediction using tree-based prediction models", *Int. J. Intell. Syst. Appl. Eng.,* vol. 11, no. 1s, pp. 96-102, 2023.

[11] R. Mohan Pattanayak, H. Sekhar Behera, and S. Panigrahi, "A novel high order hesitant fuzzy time series forecasting by using mean aggregated membership value with support vector machine", *Inf. Sci.,* vol. 626, pp. 494-523, 2023. [http://dx.doi.org/10.1016/j.ins.2023.01.075]

[12] Z. Ismail, A. Yahya, and A. Shabri, "Forecasting gold prices using multiple linear regression method", *Am. J. Appl. Sci.,* vol. 6, no. 8, pp. 1509-1514, 2009.
[http://dx.doi.org/10.3844/ajassp.2009.1509.1514]

[13] M. Sinha, S. Paul, M. Ghosh, S.N. Mohanty, and R.M. Pattanayak, "Automated Lunar crater identification with chandrayaan-2 TMC-2 images using deep convolutional neural networks", *Sci. Rep.,* vol. 14, no. 1, p. 8231, 2024.
[http://dx.doi.org/10.1038/s41598-024-58438-4] [PMID: 38589419]

[14] L. Madziwa, M. Pillalamarry, and S. Chatterjee, "Gold price forecasting using multivariate stochastic model," *Resour. Policy,* vol. 76, no. August 2021, p. 102544, 2022.
[http://dx.doi.org/10.1016/j.resourpol.2021.102544]

[15] R.M. Pattanayak, H.S. Behera, and S. Panigrahi, "A novel hybrid differential evolution-psnn for fuzzy time series forecasting", *Advances in Intelligent Systems and Computing,* vol. 990, no. January, pp. 675-687, 2020.
[http://dx.doi.org/10.1007/978-981-13-8676-3_57]

[16] K.C. Sivalingam, S. Mahendran, and S. Natarajan, "Forecasting gold prices based on extreme learning machine", *Int. J. Comput. Commun. Control,* vol. 11, no. 3, pp. 372-380, 2016.
[http://dx.doi.org/10.15837/ijccc.2016.3.2009]

[17] R.M. Pattanayak, S. Panigrahi, and H.S. Behera, "High-Order Fuzzy Time Series Forecasting by Using Membership Values Along with Data and Support Vector Machine", *Arab. J. Sci. Eng.,* vol. 45, no. 12, pp. 10311-10325, 2020.
[http://dx.doi.org/10.1007/s13369-020-04721-1]

[18] S. Alsubai, A.K. Dutta, F. Alghayadh, B.H. Alamer, R.M. Pattanayak, J.V.N. Ramesh, and S.N. Mohanty, "Design of Artificial Intelligence Driven Crowd Density Analysis for Sustainable Smart Cities", *IEEE Access,* vol. 12, no. April, pp. 121983-121993, 2024.
[http://dx.doi.org/10.1109/ACCESS.2024.3390049]

[19] R.M. Pattanayak, H.S. Behera, and S. Panigrahi, "A Non-Probabilistic Neutrosophic Entropy-Based Method For High-Order Fuzzy Time-Series Forecasting", *Arab. J. Sci. Eng.,* vol. 47, no. 2, pp. 1399-1421, 2022.
[http://dx.doi.org/10.1007/s13369-021-05718-0]

[20] R.M. Pattanayak, H.S. Behera, and S. Panigrahi, "A Multi-step-Ahead Fuzzy Time Series Forecasting by Using Hybrid Chemical Reaction Optimization with Pi-Sigma Higher-Order Neural Network", *Advances in Intelligent Systems and Computing,* vol. 999, no. January, pp. 1029-1041, 2020.
[http://dx.doi.org/10.1007/978-981-13-9042-5_88]

[21] A. G. Rushikesh Ghule, "Gold price prediction using machine learning," *Int. J. Sci. Res. Eng. Manag.,* vol. 6, no. June, Mar. 2022.

[22] T. Bunnag, "The Importance of Gold and Forecasting of the Gold Price Using ARIMA , ARIMA-GARCH, and ARIMA-TGARCH Model," no. November, 2023.

[23] W. Fang, "Gold Price Forecast by Different Models", *BCP Business & Management,* vol. 36, pp. 498-506, 2023.
[http://dx.doi.org/10.54691/bcpbm.v36i.3509]

[24] K. Minal, Ghute., Mridula, S., "Efficient Machine Learning Algorithm for Future Gold Price Prediction," pp. 216–220, 2023.
[http://dx.doi.org/10.1109/ICICT57646.2023.10134197]

[25] D. Nanthiya, S. B. Gopal, S. Balakumar, M. Harisankar, and S. P. Midhun, "Gold Price Prediction using ARIMA model," *ViTECoN 2023 - 2nd IEEE Int. Conf. Vis. Towar. Emerg. Trends Commun. Netw. Technol. Proc.,* pp. 1–6, 2023.
[http://dx.doi.org/10.1109/ViTECoN58111.2023.10157017]

[26] H.S. Dasari Siva Sankar, "Gold Prices Prediction Using Random Forest", *Int. J. Res. Appl. Sci. Eng.*

Technol., vol. 11, no. V, pp. 4261-4264, 2016.

[27] Z. Yuan, "Gold and Bitcoin Price Prediction based on KNN, XGBoost and LightGBM Model", *Highlights in Science, Engineering and Technology,* vol. 39, pp. 720-725, 2023. [http://dx.doi.org/10.54097/hset.v39i.6635]

[28] D. Sarvaiya, and D. Ramchandani, "Time series analysis and forecasting of gold price using ARIMA and LSTM model", *Int. J. Res. Appl. Sci. Eng. Technol.,* vol. 10, no. 9, pp. 168-173, 2022. [http://dx.doi.org/10.22214/ijraset.2022.46555]

[29] A. Kumar Agarwal, S. Kumari, A. Professor, and Mt. Scholar, "Gold Price Prediction using Machine Learning," *Int. J. Trend Sci. Res. Dev.,* vol. Volume-4, no. Issue-5, 2020, Accessed: Jul. 01, 2024. [Online]. Available from: www.ijtsrd.com/papers/ijtsrd33143.pdf

[30] T. Tobin and R. Kashef, "Efficient prediction of gold prices using hybrid deep learning," in Lecture Notes in Computer Science (including subseries Lecture Notes in Artificial Intelligence and Lecture Notes in Bioinformatics), 2020. [http://dx.doi.org/10.1007/978-3-030-50516-5_11]

[31] P. K. S. and L. R. S. Goel, "M. Saxena, "gold and silver price prediction using hybrid machine learning models", *2022 Seventh International Conference on Parallel, Distributed and Grid Computing (PDGC),* IEEE: Solan, Himachal Pradesh, India, pp. 390-395, 2022. [http://dx.doi.org/10.1109/PDGC56933.2022.10053278]

SUBJECT INDEX

Suneeta Satpathy, Sachi Nandan Mohanty & Subhendu Kumar Pani (Eds.)

D

E

F

G

H

I

J

K

L

M

N

O

P

Q

R

S

T

U

V

W

Y

www.ingramcontent.com/pod-product-compliance
Lightning Source LLC
LaVergne TN
LVHW070950210926
848482LV00002B/165

* 9 7 9 8 8 9 8 8 1 2 1 2 6 *